SET THEORY WITH APPLICATIONS

SET THEORY WITH APPLICATIONS

Shwu-Yeng T. Lin

You-Feng Lin

Department of Mathematics
University of South Florida
Tampa, Florida

BOOK PUBLISHERS, INC.
Tampa, Florida

Printed and Published by
BOOK PUBLISHERS, INC.
P.O. Box 21492, Tampa, Florida 33622

Printed in the United States of America

Library of Congress Cataloging in Publication Data

Lin, Shwu-Yeng T
Set theory with applications

Previous ed. [c1981] published under title:
Set theory with applications.
Bibliography:
Includes index
1. Set theory. I. Lin, You-Feng, 1932 - joint author
II. Title
QA248.L54 1985 511.3'22 84-073385
ISBN 0-931541-04-2

Preface

It has been over ten years since the appearance of *Set Theory: An Intuitive Approach*, the predecessor of this book. During these years much has happened in mathematics curricula, and the demand for new math has widened. Here at the University of South Florida, the classes of set theory, traditionally attended by mathematics and mathematics education majors, now draw half of their students from the College of Engineering, especially electrical engineering and computer science majors. Because of this, we decided to include a new chapter on Boolean algebra with applications to logic circuits and digital circuit designs. This new chapter offers practical applications of the early chapters on logic and sets. Since no engineering background is required, we hope that the nonengineering reader too will enjoy seeing the applications of a subject that is as pure as set theory in our modern technology.

Presenting set theory within the framework of an axiomatic system is, without doubt, more satisfactory when one intends to be rigorous. But there is no axiomatization of set theory that is simple and easy enough for the beginning student. Therefore, we have utilized, cautiously we hope, an intuitive approach to the subject.

In addition, we have endeavored to make the book self-contained. We have included the necessary basic material for the student who wishes to go on to such courses as topology, analysis, modern algebra, etc. But it is our hope that the material in this book will prove interesting to the student and teacher in its own right. It would be unfortunate to discover that a subject as interesting as set theory be regarded as useful only as a prerequisite course.

The prerequisites for the study of this book include some contact with college mathematics but not necessarily great mathematical skill. Ideally, the reader will have taken one quarter or one semester of calculus. The book contains enough material for a one-semester or two-quarter course in introductory set theory for the undergraduate. It can also be used in a one-quarter or short summer course by omitting the last few chapters. We suggest that the instructor go faster in the beginning two chapters and proceed more slowly later in the course.

Various minor errors and obscurities in the earlier edition have been corrected, and exercise problems have been expanded to cover a wider range of difficulty and to provide greater variety.

Our hearty thanks go to a number of colleagues who took the trouble to write kind words, point out errors, and suggest revisions. These letters were very encouraging and very helpful in the preparation of the revised edition.

We would like to thank Luke Lin for reading the chapter on Boolean algebra with applications and making many valuable suggestions and Winston Lin for proofreading parts of the manuscript.

Temple Terrace, Florida

Shwu-Yeng T. Lin
You-Feng Lin

Contents

Chapter One

Elementary Logic

In this chapter we introduce a minimal amount of logic, just enough to serve as a working tool for the remainder of this book.

1. STATEMENTS AND THEIR CONNECTIVES

The study of logic is the study of the principles and methods used in distinguishing valid arguments from those that are not valid. The purpose of this preliminary chapter in logic is to help the reader to understand the principles and methods used in each step of a proof.

The starting point in logic is the term "statement," which is used in a technical sense. By a *statement* we mean a declarative sentence that is either true or false, but not both simultaneously. It is not necessary that we know whether the statement is true or false; the only qualification is that it should definitely be one or the other. Usually we can determine immediately whether a statement is true or false, but in some cases a little effort is required, and in other cases it may be impossible to reach a conclusion. The following examples should show what we mean.

EXAMPLE 1. Each of the following is a statement.

(a) Tampa is a city in the state of Florida.
(b) $2 + 1$ is 5.
(c) The digit in the 105th decimal place in the decimal expansion of $\sqrt{3}$ is 7.
(d) The moon is made of blue cheese.
(e) There is no intelligent life on Mars.

(f) It is raining.

Clearly (a) is true while (b) and (d) are false. We are in doubt about the status (true or false) of (c) and (e), but this is due only to a defect in our knowledge. Thus (c) and (e) are also statements. The truth or falsity of sentence (f) depends on the weather at the *time* this statement is made.

EXAMPLE 2. None of the following is a statement because it makes no sense to ask if any of them is true or false.

(a) Come to our party!
(b) The sky is rich.
(c) How are you today?
(d) Goodbye, honey.

The statements quoted in Example 1 are all *simple statements.* A combinations of two or more simple statements is a *compound statement.* For example, "2 + 1 is 5 *and* the digit in the 105th decimal place in the decimal expansion of $\sqrt{3}$ is 7" is a compound statement.

We are familiar with the use of letters to represent numbers in algebra. In the study of logic we use letters such as $p, q, r, \ldots$ to represent statements. A letter such as p may represent either a simple statement or a compound statement. Unless otherwise stated, we shall use capital letters $P, Q, R, \ldots$ to represent compound statements.

There are many ways of connecting statements such as $p, q, r, \ldots$ to form compound statements, but only five are used frequently. These five common *connectives* are (a) "not," symbolized by $\sim$; (b) "and," symbolized by $\wedge$; (c) "or," symbolized by $\vee$; (d) "if . . . then . . ." symbolized by $\rightarrow$; and (e) ". . . if and only if . . ." symbolized by $\leftrightarrow$.

In this section we discuss the connectives $\sim$ and $\wedge$, postponing the remaining connectives $\vee$, $\rightarrow$, and $\leftrightarrow$ until the next section.

Let p be a statement. Then the statement $\sim p$, read "*not p*" or "the *negation* of p," is true whenever the statment p is false and is false whenever p is true. For example, let p be the statement "This is an easy course." Then its negation $\sim p$ represents "This is not an easy course."

The truth of $\sim p$ depends upon the truth of p. It is convenient to record this dependency in a *truth table*:

Table 1

p	$\sim p$
T	F
F	T

where the letters T and F stand for "true" and "false," respectively. In the first column of Table 1 we list the two possible truth values for the statement p, namely T or F. Each line in a truth table represents a case that must be considered, and of course in this very simple situation, there are only two cases. Using the lines in Table 1 we see that if p is true then $\sim p$ is false, and if p is false, then $\sim p$ is true. Thus Table 1 tells us the truth value of $\sim p$ in every case.

Definition 1. The connective $\wedge$ may be placed between any two statements p and q to form the compound statement $p \wedge q$ whose truth values are given in the following truth table.

Table 2

p	q	$p \wedge q$
T	T	T
T	F	F
F	T	F
F	F	F

The symbol $p \wedge q$ is read "*p and q*" or the "*conjunction* of p and q." For example, let p be the statement "The sky is blue" and q be "The roses are red." In a compound statement such as $p \wedge q$, the individual statements p and q are called *components.* A component may be a simple statement or a compound statement. In a compound statement with two components, such as $p \wedge q$, there are at most $4(=2 \times 2)$ possibilities, called the *logical possibilities*, to be considered; namely:

(1) p is true and q is true;
(2) p is true and q is false;
(3) p is false and q is true;
(4) p is false and q is false;

Each one of these four possibilities is covered in the four rows of Table 2. The last column gives the truth values of $p \wedge q$. Inspection shows that $p \wedge q$ is true in only one case. That is, $p \wedge q$ is true when both components are true, and in the other three cases $p \wedge q$ is false. The thoughtful reader will realize that Table 2 reflects the way in which the conjunction "and" is used in everyday English.

Using Tables 1 and 2, we can find the truth values of complicated statements involving only the connectives $\sim$ and $\wedge$.

EXAMPLE 3. Construct the truth table for the compound statement

$$\sim[(\sim p)\wedge(\sim q)]$$

Solution.

Table 3

p	q	$\sim p$	$\sim q$	$(\sim p)\wedge(\sim q)$	$\sim[(\sim p)\wedge(\sim q)]$
T	T	F	F	F	T
T	F	F	T	F	T
F	T	T	F	F	T
F	F	T	T	T	F
Step		1	1	2	3

If the method used in constructing Table 3 is not obvious, a word of explanation may help. The headings are selected so that the compound statement (last column) is gradually built from its various components. The first two columns merely record all cases for the truth values of p and q. We then use Table 1 to obtain the entries in the third and fourth columns, the corresponding truth values for $\sim p$ and $\sim q$. In the next step we use the entries in the third and fourth columns and Table 2 to obtain the entries in the fifth column. finally the entries in the fifth column and Table 1 give the entries in the sixth column—the truth values of $\sim[(\sim p)\wedge(\sim q)]$. The serious student should now copy this last compound statement, close the book, and try to reproduce Table 3.

The statement in the above example, $\sim[(\sim p)\wedge(\sim q)]$, uses parentheses and brackets to indicate the order in which the connectives apply. Often an expression can be simplified if we can drop some of the parentheses or brackets. The usual convention is to agree that $\sim$ binds more strongly than $\wedge$, i.e., the connective $\sim$ should be applied first. Thus, for example, the expression $(\sim p)\wedge(\sim q)$ is simplified as $\sim p\wedge\sim q$.

EXERCISE 1.1

In problems 1 through 10 an English sentence is given. In each case determine whether the sentence is a statement (S), or not (N).

1. On January 7, 1442, snow fell somewhere in Florida.
2. Aristotle had flat feet.
3. Socialism is wrong.
4. The richest man in the world is Mr. Hunt of Texas.

5. Jack and Jill are good.
6. How much is this car worth?
7. Keep off the grass.
8. Always fasten your seat belt.
9. The number $2^{987654321} + 37$ is prime.
10. Beethoven wrote some of Chopin's music.
11. Among the statements given in Problems 1 through 10, indicate those that you feel must be true (T), and those for which the status may be difficult to determine (D).

In problems 12 through 19 find the truth values for the given statements. Use the format of Table 1 or Table 2 for the two or four cases respectively.

12. $\sim(\sim p)$
13. $\sim[\sim(\sim p)]$
14. $p \wedge p$
15. $\sim(p \wedge \sim p)$
16. $p \wedge \sim q$
17. $\sim p \wedge q$
18. $(p \wedge q) \wedge \sim p$
19. $\sim(p \wedge q)$
20. In a compound statement involving three distinct components p, q, and r, how many cases are required to cover every logical possibility? How many cases are necessary if there are four distinct components? How many cases are necessary if there are n distinct components?
21. The following is an attempt to arrange all the cases in a truth table for a statement involving three components p, q, and r. Complete this unfinished work.

p	q	r	
T	T		
T	T	F	
T		T	
T		F	
	T	T	
	T	F	
F	F		
	F	F	

In Problems 22 through 25, find the truth tables for the given statements. Use the pattern developed in Problem 21 for the various cases.

22. $(p \wedge q) \wedge r$
23. $p \wedge (q \wedge r)$
24. $(p \wedge \sim q) \wedge r$
25. $\sim q \wedge (r \wedge p)$

2. THREE MORE CONNECTIVES

In the English language there is an ambiguity involved in the use of "or." The statement "I will get a Master's degree or a Ph.D.," indicates that the speaker may get both the Master's degree and the Ph.D. But in another statement, "I will marry Linda or Lucy," the word "or" means that only one of the two girls will be chosen. In mathematics and logic, we cannot allow ambiguity. Hence we must decide on the meaning of the word "or."

Definition 2. The connective $\vee$ may be placed between any two statements p and q to form the compound statement $p \vee q$. The truth values of $p \vee q$ are defined by Table 4. Thus $\vee$ is defined to be the inclusive "or" as used in the first statement above.

Table 4

p	q	$p \vee q$
T	T	T
T	F	T
F	T	T
F	F	F

The symbol $p \vee q$ is read "p or q" or "the *disjunction* of p and q." Notice that the conjunction of p and q is true when and only when the two components are both true (Table 2), whereas the disjunction is false when and only when the two components are false (Table 4).

Let us compare the truth tables for $p \vee q$ and $\sim(\sim p \wedge \sim q)$ in Tables 3 and 4. We find that in each case the last column is $TTTF$ so that these two statements have the same truth values in each of the four logical possibilities. Showing that certain statements have the same truth values in each case is an important part of logic. In fact, logic treats two such statements as the same.

Definition 3. When two statements P and Q, simple or compound, have the same truth values in each of all the logical possibilities, then P is said to be *logically equivalent* or simply *equivalent* to Q, and we write $P \equiv Q$.

In short, two statements are logically equivalent provided that they have the same truth table. Thus, we have

$$p \vee q \equiv \sim(\sim p \wedge \sim q)$$

Although two logically equivalent statements are regarded as the same, as far as logic is concerned, we prefer the simpler statement "p or q" to its more complicated equivalent statement. "It is not true that neither p nor q."

Definition 4. The connective $\rightarrow$ is called the *conditional* and may be placed between *any* two statements p and q to form the compound statement $p \rightarrow q$ (read: "if p then q"). By definition the statement $p \rightarrow q$ is equivalent to the statement $\sim(p \wedge \sim q)$, and the truth values of $p \rightarrow q$ are given in Table 5.

Table 5

Case	p	q	$\sim q$	$p \wedge \sim q$	$p \rightarrow q[\equiv \sim(p \wedge \sim q)]$
1	*T*	*T*	*F*	*F*	*T*
2	*T*	*F*	*T*	*T*	*F*
3	*F*	*T*	*F*	*F*	*T*
4	*F*	*F*	*T*	*F*	*T*

A motivation of Definition 4 is in order. Let p be the statement "It is raining" and let q be the statement "I will give you a ride." Then the compound statement $p \rightarrow q$ represents "If it is raining then I will give you a ride." When do we consider such a contract (statement) breached (false)? Clearly, the contract ($p \rightarrow q$) is breached if, and only if it is raining (p is true), and I am not giving you a ride (q is false), and only in this case. That is, the final column of the truth table of $p \rightarrow q$ must have the value "*T*" in every case, except case 2.

According to Definition 4, the meaning of the conditional statement $p \rightarrow q$ departs radically from our ordinary usage of "If p then q." In our ordinary language,[1] a sentence of the form "If p then q" is taken to mean that q is true whenever p is true. Therefore the cases in which p is false need not be considered.

For example, the statement, "If Lincoln shot Grant, then Jefferson was the first president" is regarded as nonsense, since both of the components are false. Consequently in ordinary usage one does not inquire whether such a compound statement is true. In creating a formal language, the logi-

[1] As opposed to "ordinary language" logic is called a formal language.

cian wishes to assign a truth value to $p \to q$ for each of the four logical possibilities, even though two of the cases appear to be nonsense in our ordinary language. For a variety of reasons, which will appear in due time, logicians have settled on the definition adopted here. Thus in our formal language $p \to q$ is true in every case except case 2 (see Table 5). As a consequence of this agreement, we will be able to prove some very simple and useful theorems, without such an agreement the proofs of which would be awkward or very difficult.[2]

We now introduce the last of the five most common connectives, one that appears frequently in the statements of mathematical theorems.

Definition 5. The connective $\leftrightarrow$ is called the *biconditional* and may be placed between any two statements p and q to form the compound statement $p \leftrightarrow q$ (read: "*p if and only if q*"). The statement $p \leftrightarrow q$ is defined to be equivalent to the compound statement $(p \to q) \wedge (q \to p)$, and the truth values of $p \leftrightarrow q$ are given in Table 6.

EXAMPLE 4. Find the truth table for $p \leftrightarrow q$.

Solution. Following the method described earlier we obtain Table 6.

Table 6

Case	p	q	$p \to q$	$q \to p$	$p \leftrightarrow q [\equiv (p \to q) \wedge (q \to p)]$
1	*T*	*T*	*T*	*T*	*T*
2	*T*	*F*	*F*	*T*	*F*
3	*F*	*T*	*T*	*F*	*F*
4	*F*	*F*	*T*	*T*	*T*
Step			1	1	2

From the above truth table, we observe that $p \leftrightarrow q$ is true if both components are true or if both components are false. In any other case (case 2 and 3) the statement $p \leftrightarrow q$ is false.

[2] See, for instance, Theorems 1 and 7 in Chapter 2.

EXERCISE 1.2

In problems 1 through 12 construct the truth tables for the given statements.

1. $p \vee \sim p$
2. $\sim(p \vee \sim p)$
3. $\sim(\sim p \vee \sim p)$
4. $\sim p \vee q$
5. $(\sim q) \rightarrow (\sim p)$
6. $q \leftrightarrow p$
7. $p \wedge (q \vee r)$
8. $(p \wedge q) \vee (p \wedge r)$
9. $p \vee (q \wedge r)$
10. $(p \vee q) \wedge (p \vee r)$
11. $(p \vee q) \vee r$
12. $p \vee (q \vee r)$
13. Is the statement $(\sim q) \rightarrow (\sim p)$ (Problem 5) logically equivalent to the statement $p \rightarrow q$?
14. Is the statement $\sim p \vee q$ (Problem 4) logically equivalent to the statement $p \rightarrow q$?
15. From the statements in Problems 1 through 12, find those pairs of statements that are logically equivalent.
16. In each of the following, translate the given compound statement into a symbolic form using the suggested symbols.
    (a) It is not the case that I am not friendly to you. (F)
    (b) If she is an angel, then she has two wings. (A, W)
    (c) The price of meat goes up if and only if the supply does not meet the demand of meat. (P, S)
    (d) Either the farmers will reduce the prices or the government will step in. (F, G)
    (e) If meat exports increase or more livestock is not raised, then prices will rise. (E, R, P)
17. Sometimes $p \rightarrow q$ is read "p only if q." Translate each of the following statements into a symbolic form using the suggested symbols.
    (a) I will pass this course only if I work hard. (P, W)
    (b) A function is differentiable only if it is continuous. (D, C)
    (c) A matrix has an inverse only if its determinant is not zero. (I, Z)
    (d) A matrix has a nonzero determinant only if it has an inverse. (Z, I)
18. In the statement $p \rightarrow q$, p is also called *sufficient condition* for q, and q is called a *necessary condition* for p. Rephrase the statements (a) to (d) of 17, above, by first using the language of "sufficient condition" and then using the language of "necessary condition."
19. Is $\sim(p \leftrightarrow q)$ logically equivalent to $p \leftrightarrow \sim q$? Prove your conclusion by comparing the truth tables.
20. Is $p \leftrightarrow \sim q$ logically equivalent to $\sim p \leftrightarrow q$?
21. If a compound statement involving simple statements p and q is known to have the following truth table:

p	q	?
T	T	F
T	F	F
F	T	F
F	F	T

Can you find such a compound statement?

3. TAUTOLOGY, IMPLICATION, AND EQUIVALENCE

Let us examine the truth table for the statement $p \vee \sim p$:

Table 7

p	$\sim p$	$p \vee \sim p$
T	F	T
F	T	T

We notice that the statement $p \vee \sim p$ is true in every case, that is, in all logical possibilities. Such an important type of statement deserves a special name.

Definition 6. A statement is said to be a *tautology* provided that it is true in each of all logical possibilities.

Let P and Q be two statement, compound or simple. If the conditional statement $P \rightarrow Q$ is a tautology, it is called an *implication* and is denoted by $P \Rightarrow Q$ (read: P *implies* Q). Thus, the following conditional statements are implications:

(1) $p \rightarrow p$.
(2) $p \wedge q \rightarrow q \wedge p$.
(3) $p \rightarrow p \wedge p$.
(4) $p \wedge q \rightarrow q$.[3]

In logic or mathematics, "theorems" are meant to be true statements, and a "proof" (of a theorem) is a justification of the theorem.

[3] We shall consider $\vee$ and $\wedge$ to bind more strongly than $\rightarrow$ and $\leftrightarrow$, and shall write $p \rightarrow p \vee q$ for $p \rightarrow (p \vee q)$, etc. See also the last paragraph of Section 1.

Theorem 1. Let p and q be any two statements. Then

(a) Law of Addition (Add.): $p \Rightarrow p \vee q$.

(b) Laws of Simplification (Simp.): $p \wedge q \Rightarrow p$,
$p \wedge q \Rightarrow q$.

(c) Disjunctive Syllogism (D.S.): $(p \vee q) \wedge \sim p \Rightarrow q$.

Proof. We leave the proofs for (a) and (b) to the reader as exercises.
The following is a simplified truth table for $(p \vee q) \wedge \sim p \rightarrow q$:

Table 8

	$(p$	$\vee$	$q)$	$\wedge$	$\sim p$	$\rightarrow$	q
	T	T	T	F	F	T	T
	T	T	F	F	F	T	F
	F	T	T	T	T	T	T
	F	F	F	F	T	T	F
Step	1	2	1	3	2	4	1

Let us take a moment to explain the construction of a simplified truth table: The truth values in Table 8 are assigned, column by column, in the order indicated by the numerals appearing in the bottom row of the table. In a simplified truth table, we write the truth values directly first under each component and then under the connectives. This saves space and time.

Now, returning to the proof of the theorem, since the final step (step 4) in Table 8 consists of all T's, the conditional statement $(p \vee q) \wedge \sim p \rightarrow q$ is indeed an implication.

If the biconditional statement $P \leftrightarrow Q$ happens to be a tautology, it is called an *equivalence* and is denoted by $P \Leftrightarrow Q$ (read: P is equivalent to Q).

From Definition 5 and Table 6, $P \Leftrightarrow Q$ provided that P and Q have the same truth values in each of all logical possibilities, and conversely, P and Q have the same truth values in each of all logical possibilities provided $P \Leftrightarrow Q$. Thus, by Definition 3, $P \Leftrightarrow Q$ and $P \equiv Q$ have the same meaning, and hence we may use $\Leftrightarrow$ and $\equiv$ interchangeably.

Theorem 2. Let p and q be any two statements. Then

(a) Law of Double Negation (D.N.): $\sim(\sim p) \equiv p$.

(b) Commutative Laws (Com.): $p \wedge q \equiv q \wedge p$,
$p \vee q \equiv q \vee p$.

(c) Laws of Idempotency (Idemp.): $p \wedge p \equiv p$,
$p \vee p \equiv p$.

(d) Contrapositive Law (Contrap.): $(p \rightarrow q) \equiv (\sim q \rightarrow \sim p)$.

Proof. We leave proofs for parts (a), (b), and (c) to the reader as exercises, but we give an outline of the proof of (d).

We have the following simplified truth table for the biconditional statement $(p \to q) \leftrightarrow (\sim q \to \sim p)$:

Table 9

	$(p$	$\to$	$q)$	$\leftrightarrow$	$(\sim q$	$\to$	$\sim p)$
	T	T	T	T	F	T	F
	T	F	F	T	T	F	F
	F	T	T	T	F	T	T
	F	T	F	T	T	T	T
Step	1	2	1	4	2	3	2

Thus, Table 9 shows that $p \to q$ is equivalent to $\sim q \to \sim p$.

The following theorem credited to Augustus De Morgan (1806-1871) is one of the most convenient tools in logic.

Theorem 3. (*De Morgan's Laws* (*De M.*)). Let p and q be any two statements. Then $\sim (p \wedge q) \equiv \sim p \vee \sim q$
and $\sim (p \vee q) \equiv \sim p \wedge \sim q$.

Proof. We shall prove the first part of this theorem and leave the other part to the reader as an exercise. We construct a simplified truth table for the biconditional $\sim (p \wedge q) \leftrightarrow (\sim p \vee \sim q)$:

Table 10

	$\sim$	$(p$	$\wedge$	$q)$	$\leftrightarrow$	$(\sim p$	$\vee$	$\sim q)$
	F	T	T	T	T	F	F	F
	T	T	F	F	T	F	T	T
	T	F	F	T	T	T	T	F
	T	F	F	F	T	T	T	T
Step	3	1	2	1	4	2	3	2

The above truth table shows that $\sim (p \wedge q)$ is equivalent to $\sim p \vee \sim q$.

Theorem 4. Let p, q, and r be any statements. Then

(a) Associative Laws (Assoc.): $(p \wedge q) \wedge r \equiv p \wedge (q \wedge r)$,
$(p \vee q) \vee r \equiv p \vee (q \vee r)$.

(b) Distributive Laws (Dist.): $p \wedge (q \vee r) \equiv (p \wedge q) \vee (p \wedge r)$,
$p \vee (q \wedge r) \equiv (p \vee q) \wedge (p \vee r)$.

(c) Transitive Law (Trans.): $(p \rightarrow q) \wedge (q \rightarrow r) \Rightarrow (p \rightarrow r)$.

Proof. We leave the proofs for the Associative Laws and the second Distributive Law to the reader as exercises.

Let us prove that $p \wedge (q \vee r) \equiv (p \wedge q) \vee (p \wedge r)$. Since this involves three components, there are $2^3 = 8$ logical possibilities to consider. The following truth table shows that $p \wedge (q \vee r)$ and $(p \wedge q) \vee (p \wedge r)$ have the same truth values in each of all eight logical possibilities. Therefore, $p \wedge (q \vee r)$ and $(p \wedge q) \vee (p \wedge r)$ are equivalent.

Table 11

p	q	r	$q \vee r$	$p \wedge q$	$p \wedge r$	$p \wedge (q \vee r)$	$(p \wedge q) \vee (p \wedge r)$
T	T	T	T	T	T	T	T
T	T	F	T	T	F	T	T
T	F	T	T	F	T	T	T
T	F	F	F	F	F	F	F
F	T	T	T	F	F	F	F
F	T	F	T	F	F	F	F
F	F	T	T	F	F	F	F
F	F	F	F	F	F	F	F

For simplicity and space saving, we construct a simplified truth table as introduced in Table 8 for $(p \rightarrow q) \wedge (q \rightarrow r) \rightarrow (p \rightarrow r)$.

Table 12

	$(p$	$\rightarrow$	$q)$	$\wedge$	$(q$	$\rightarrow$	$r)$	$\rightarrow$	$(p$	$\rightarrow$	$r)$
	T	T	T	T	T	T	T	T	T	T	T
	T	T	T	F	T	F	F	T	T	F	F
	T	F	F	F	F	T	T	T	T	T	T
	T	F	F	F	F	T	F	T	T	F	F
	F	T	T	T	T	T	T	T	F	T	T
	F	T	T	F	T	F	F	T	F	T	F
	F	T	F	T	F	T	T	T	F	T	T
	F	T	F	T	F	T	F	T	F	T	F
Step	1	2	1	3	1	2	1	4	1	2	1

Since the final step (step 4) consists entirely of T values, the Transitive Law is proved.

Because of the Associative Laws, the brackets in $(p \wedge q) \wedge r \equiv p \wedge (q \wedge r)$ and $(p \vee q) \vee r \equiv p \vee (q \vee r)$ become unnecessary, and the expressions $p \wedge q \wedge r$ and $p \vee q \vee r$ now have definite meanings, and similarly for $p_1 \wedge p_2 \wedge \cdots \wedge p_n$ and $p_1 \vee p_2 \vee \cdots \vee p_n$.

Theorem 5. Let p, q, r, and s be any statements. Then

(a) *Constructive Dilemmas* (C.D.):

$$(p \rightarrow q) \wedge (r \rightarrow s) \Rightarrow (p \vee r \rightarrow q \vee s),$$
$$(p \rightarrow q) \wedge (r \rightarrow s) \Rightarrow (p \wedge r \rightarrow q \wedge s).$$

(b) *Destructive Dilemmas* (D.D.):

$$(p \rightarrow q) \wedge (r \rightarrow s) \Rightarrow (\sim q \vee \sim s \rightarrow \sim p \vee \sim r),$$
$$(p \rightarrow q) \wedge (r \rightarrow s) \Rightarrow (\sim q \wedge \sim s \rightarrow \sim p \wedge \sim r).$$

Proof. The proof of Theorem 5 is left to the reader as an exercise.

Theorem 6. Let p and q be statements. Then

(a) *Modus Ponens* (M.P.): $(p \rightarrow q) \wedge p \Rightarrow q$.
(b) *Modus Tollens* (M.T.): $(p \rightarrow q) \wedge \sim q \Rightarrow \sim p$.
(c) *Reductio ad Absurdum* (R.A.): $(p \rightarrow q) \Leftrightarrow (p \wedge \sim q \rightarrow q \wedge \sim q)$.

Proof. Exercise

EXERCISE 1.3

1. Prove parts (a) and (b) of Theorem 1.
2. Prove parts (a), (b), and (c) of Theorem 2.
3. Prove that $\sim (p \vee q) \equiv \sim p \wedge \sim q$.
4. Prove part (a) of Theorem 4.
5. Prove that $p \vee (q \wedge r) \equiv (p \vee q) \wedge (p \vee r)$.
6. Prove that $(p \rightarrow q) \Rightarrow (p \wedge r \rightarrow q \wedge r)$.
7. Prove that $(p \leftrightarrow q) \equiv (p \wedge q) \vee (\sim p \wedge \sim q)$.
8. Using De Morgan's Law, write in ordinary language the negation of the statement "This function has a derivative or I am stupid."
9. Prove the following De Morgan's Laws for three components.
   (a) $\sim (p \wedge q \wedge r) \equiv \sim p \vee \sim q \vee \sim r$
   (b) $\sim (p \vee q \vee r) \equiv \sim p \wedge \sim q \wedge \sim r$

10. Can you generalize, without proof, De Morgan's Laws for n components? See Problem 9 for $n = 3$.
11. Prove the following *Absorption Laws.*
    (a) $p \wedge (p \vee r) \equiv p$
    (b) $p \vee (p \wedge q) \equiv p$
12. Prove Theorem 5.
13. Prove Theorem 6.
14. Is $(p \vee q) \wedge (r \vee s)$ logically equivalent to $(p \wedge r) \vee (p \wedge s) \vee (q \wedge r) \vee (q \wedge s)$?
15. Is $(p \wedge q) \vee (r \wedge s)$ logically equivalent to $(p \vee r) \wedge (p \vee s) \wedge (q \vee r) \wedge (q \vee s)$?
16. Prove that $(p \vee q) \wedge \sim p$ is logically equivalent to $q \wedge \sim p$.
17. Prove that $(p \vee q) \wedge (\sim p \vee \sim q)$ is logically equivalent to $(p \wedge \sim q) \vee (\sim p \wedge q)$.

In each of problems 18 through 20, the final column of the truth table of a certain unknown statement involving simple statements p, q and r is given, find such a compound statement.

18. *TFFFFFFF.*
19. *FFFFFFFT.*
20. *TFFFFFFT.* [Hint: If X and Y are respectively solutions to problems 18 and 19, $X \vee Y$ is a solution to problem 20.]

4. CONTRADICTION

In contrast to tautologies, there are statements whose truth values are all false for each of the logical possibilities. Such statements are called *contradictions.* For example, $p \wedge \sim p$ is a contradiction.

It is obvious that if t is a tautology then $\sim t$ is a contradiction; conversely, if c is a contradiction then $\sim c$ is a tautology.

Theorem 7. Let t, c, and p be a tautology, a contradiction, and an arbitrary statement, respectively.
Then,
(a) $p \wedge t \Leftrightarrow p$,
$p \vee t \Leftrightarrow t$.
(b) $p \vee c \Leftrightarrow p$,
$p \wedge c \Leftrightarrow c$.
(c) $c \Rightarrow p$, and $p \Rightarrow t$.

Proof. (a) The following truth table for $p \wedge t \leftrightarrow p$ shows that $p \wedge t$ is equivalent to p.

Table 13

	p	$\wedge$	t	$\leftrightarrow$	p
	T	T	T	T	T
	F	F	T	T	F
Step	1	2	1	3	1

The other equivalence $p \vee t \Leftrightarrow t$ may be proved similarly.

(b) From the following truth table we find that the conditional statement $p \vee c \leftrightarrow p$ is a tautology and hence, $p \vee c \Leftrightarrow p$.

Table 14

	p	$\vee$	c	$\leftrightarrow$	p
	T	T	F	T	T
	F	F	F	T	F
Step	1	2	1	3	1

The proof for $p \wedge c \Leftrightarrow c$ is similar.

(c) The truth tables of $c \rightarrow p$ and $p \rightarrow t$ show that $c \Rightarrow p$ and $p \Rightarrow t$ are tautologies.

Table 15

c	$\rightarrow$	p
F	T	T
F	T	F

p	$\rightarrow$	t
T	T	T
F	T	T

For the remainder of this book, *the symbol c with or without a subscript will stand for a contradiction; and the symbol t with or without a subscript will denote a tautology.*

EXERCISE 1.4

1. Prove that $p \vee t \Leftrightarrow t$ and $p \wedge c \Leftrightarrow c$.
2. Prove that $\sim t \Leftrightarrow c$ and $\sim c \Leftrightarrow t$.
3. Prove the following Reductio ad Absurdum.

$$(p \wedge \sim q \rightarrow c) \Leftrightarrow (p \rightarrow q)$$

4. Prove that $p \wedge (p \rightarrow q) \wedge (p \rightarrow \sim q) \Leftrightarrow c$.
5. Prove that $(p \rightarrow q) \Rightarrow (p \vee r \rightarrow q \vee r)$ for any statement r.

6. Joe claimed he could do anything. Could Joe make an object that he could not lift?

5. DEDUCTIVE REASONING

The 16 laws summarized in Theorems 1 through 6 are very useful tools for justifying logical equivalences and implications, as illustrated in Examples 5 through 7. We shall call these 16 laws the *rules of inference.* It should be noted that these rules are selected just for convenient references and are not intended to be independent of each other. For instance, the Contrapositive Law can be established "deductively" by using other laws and relevant definitions, as the next example shows.

EXAMPLE 5. Prove the Contrapositive Law, $(p \rightarrow q) \equiv (\sim q \rightarrow \sim p)$, by using relevant definitions and other rules of inference.

Solution.

$$\begin{aligned} (p \rightarrow q) &\equiv \sim(p \wedge \sim q) && \text{Def. 4} \\ &\equiv \sim(\sim q \wedge p) && \text{Com.} \\ &\equiv \sim[\sim q \wedge \sim(\sim p)] && \text{D.N.} \\ &\equiv (\sim q \rightarrow \sim p) && \text{Def. 4} \end{aligned}$$

Therefore, $(p \rightarrow q) \equiv (\sim q \rightarrow \sim p)$, by the Transitive Law.

The method of proof used in Example 5 is called *deductive reasoning* or the *deductive method*, which differs from the method of proof by truth tables.

In general, in deductive reasoning any previously stated axioms, definitions, and theorems and the rules of inference may be used.

EXAMPLE 6. Prove the Disjunctive Syllogism by deductive reasoning.

Solution.

$$\begin{aligned} (p \vee q) \wedge \sim p &\equiv \sim p \wedge (p \vee q) && \text{Com.} \\ &\equiv (\sim p \wedge p) \vee (\sim p \wedge q) && \text{Dist.} \\ &\equiv c \vee (\sim p \wedge q) && \sim p \wedge p \equiv c \\ &\equiv (\sim p \wedge q) \vee c && \text{Com.} \\ &\equiv \sim p \wedge q && \text{Th. 7(b)} \\ &\Rightarrow q && \text{Simp.} \end{aligned}$$

Finally, by the Transitive Law, $(p \vee q) \wedge \sim p \Rightarrow q$.

EXAMPLE 7. Prove the following *Exportation Law*:

$$(p \wedge q \rightarrow r) \equiv [p \rightarrow (q \rightarrow r)]$$

by deductive reasoning.

Solution.

$$\begin{aligned} [p \to (q \to r)] &\equiv [p \to \sim(q \wedge \sim r)] && \text{Def. 4} \\ &\equiv \sim[p \wedge (q \wedge \sim r)] && \text{Def. 4, D.N.} \\ &\equiv \sim[(p \wedge q) \wedge \sim r] && \text{Assoc.} \\ &\equiv (p \wedge q \to r) && \text{Def. 4} \end{aligned}$$

Hence, $(p \wedge q \to r) \equiv [p \to (q \to r)]$.

EXAMPLE 8. Prove that $(p \to r) \vee (q \to s) \equiv (p \wedge q \to r \vee s)$ by deductive reasoning.

Solution.

$$\begin{aligned} (p \to r) \vee (q \to s) &\equiv \sim(p \wedge \sim r) \vee \sim(q \wedge \sim s) && \text{Def. 4} \\ &\equiv (\sim p \vee r) \vee (\sim q \vee s) && \text{De M., D.N.} \\ &\equiv (\sim p \vee \sim q) \vee (r \vee s) && \text{Com., Assoc.} \\ &\equiv \sim[(p \wedge q) \wedge \sim(r \vee s)] && \text{De M., D.N.} \\ &\equiv (p \wedge q \to r \vee s) && \text{Def. 4} \end{aligned}$$

Why we want to use deductive reasoning as opposed to truth tables may be seen from the following comparison: To verify the equivalence in Example 8 by the method of truth tables, we would have to construct a huge truth table with 16 $(=2^4)$ cases (see Problem 20 of Exercise 1.1 or Problem 12 of Exercise 1.3); on the other hand, in the solution of Example 8, above, we established that equivalence in only five short steps.

EXERCISE 1.5

Prove the following tautologies by the deductive method.

1. Modus Ponens: $p \wedge (p \to q) \Rightarrow q$
2. Modus Tollens: $\sim q \wedge (p \to q) \Rightarrow \sim p$
3. Reductio ad Absurdum: $(p \to q) \Leftrightarrow (p \wedge \sim q \to c)$
4. Disjuctive Syllogism: $(p \vee q \vee r) \wedge \sim r \wedge \sim q \Rightarrow p$
5. Theorem 7(c): $c \Rightarrow p$
6. $(p \to q) \Leftrightarrow (p \to p \wedge q)$
7. $(p \to q) \Leftrightarrow (p \vee q \to q)$
8. $(p \to q) \Leftrightarrow \sim p \vee q$
9. $(p \to r) \wedge (q \to r) \Leftrightarrow (p \vee q \to r)$
10. $(p \to q) \wedge (p \to r) \Leftrightarrow (p \to q \wedge r)$
11. $(p \to q) \wedge (p \to \sim q) \Leftrightarrow \sim p$
12. $(p \to q) \vee (p \to r) \Leftrightarrow (p \to q \vee r)$
13. $(p \to r) \vee (q \to r) \Leftrightarrow (p \wedge q \to r)$
14. $(p \to q) \wedge (p \vee q) \Leftrightarrow q$
15. $(p \to q \wedge r) \Rightarrow (p \to r)$
16. $[p \to (q \to r)] \wedge (p \to q) \Rightarrow (p \to r)$
17. $(p \to q) \wedge (q \to r) \wedge p \Rightarrow r$
18. $[(p \to q) \vee (p \to r)] \wedge p \Rightarrow q \vee r$

19. $(p \to q) \wedge (p \to r) \wedge p \Rightarrow q \vee r$
20. $(p \to q) \wedge (r \to s) \wedge (\sim q \vee \sim s) \Rightarrow \sim p \vee \sim r$

6. QUANTIFICATION RULES

In any general discussion, we have in mind a particular *universe* or *domain of discourse*, that is, a collection of objects whose properties are under consideration. For example, in the statement "All humans are mortal," the universe is the collection of all humans. With this understanding of the universe, the statement "All humans are mortal" may be alternatively expressed as:

For all x in the universe, x is mortal.

The phrase "For all x in the universe" is called a *universal quantifier*, and is symbolized as $(\forall x)$. The sentence "x is mortal" says something about x; we symbolize this as $p(x)$. Using these new symbols we can now write the general statement "All humans are mortal" as

$$(\forall x)\ (p(x))$$

Next, consider the statement "Some humans are mortal." Here the universe (or domain of discourse) is still the same as in the previous statement. With this universe in mind, we can restate the statement "Some humans are mortal" successively as:

There exists at least one individual who is mortal.
There exists at least one x such that x is mortal.

and as

There exists at least one x such that $p(x)$.

The phrase "There exists at least one x such that" is called an *existential quantifier* and is symbolized as $(\exists x)$. Using this new symbol we can now rewrite the statement "some humans are mortal" as

$$(\exists x)\ (p(x))$$

In general, suppose we have a domain of discourse U and a general statement $p(x)$, called a *propositional predicate*, whose "variable" x ranges over U. Then $(\forall x)\ (p(x))$ asserts that for all x in U, the statement $p(x)$ about x is true, and $(\exists x)\ (p(x))$ means that there exists at least one x in U such that $p(x)$ is true.

In elementary mathematics, quantifiers are often suppressed for the sake of simplicity. For example, "$(x + 1)(x - 1) = x^2 - 1$" in high school algebra books should be understood to say that "for every real number x, $(x + 1)(x - 1) = x^2 - 1$." In mathematics, "for every" and "for all" mean the same and both are symbolized by $\forall$; and "for some" means the same as "there exists" and is symbolized by $\exists$. In less formal expressions, we often put the quantifier after the statement. For example, the statement "$f(x) = 0$ for all x" is just the same as "$(\forall x)(f(x) = 0)$."

In logic and in mathematics, the negation of the statement "$p(x)$ is true for every x (in U)," $\sim[(\forall x)(p(x))]$, is considered to be the same as the assertion "there exists at least one x (in U) for which $p(x)$ is false," $(\exists x)(\sim p(x))$. Similarly, $\sim[(\exists x)[p(x))]$ is considered the same as "there is no x (in U) such that $p(x)$ is true"; or in other words, "$p(x)$ is false for all x (in U), " or $(\forall x)(\sim p(x))$. We summarize in the following

Rule of Quantifier Negation (Q.N.). Let $p(x)$ be a propositional predicate, that is, a statement about an unspecified object x in a given universe. Then,

$$\sim[(\forall x)(p(x))] \equiv (\exists x)(\sim p(x))$$

and

$$\sim[(\exists x)(p(x))] \equiv (\forall x)(\sim p(x))$$

We have used "$\equiv$" to denote that two quantified statements on both sides of $\equiv$ are considered the same in logic; this usage is consistent with the usage of $\equiv$ for logical equivalences, as will be seen in the next paragraph.

To better understand the quantified statements $(\forall x)(p(x))$ and $(\exists x)(p(x))$, let us inspect the case in which the universe of discourse consists of finitely many individuals denoted by $a_1, a_2, a_3, \ldots a_n$. Then, since $(\forall x)(p(x))$ asserts that $p(x)$ is true for every $a_1, a_2, a_3, \ldots a_n$, the statement $(\forall x)(p(x))$ is true if and only if the conjunction of

$$p(a_1),\ p(a_2),\ p(a_3), \ldots,\ p(a_n)$$

is true. Thus,

$$(\forall x)(p(x)) \text{ amounts to } p(a_1) \wedge p(a_2) \wedge \cdots \wedge p(a_n)$$

Similarly,

$$(\exists x)(p(x)) \text{ means } p(a_1) \vee p(a_2) \vee \cdots \vee p(a_n)$$

Thus, the Rule of Quantifier Negation may be viewed as a generalization of De Morgan's Laws (Theorem 3).

EXAMPLE 9. Which of the following is equivalent to the negation of the statement "All snakes are poisonous?"

(a) All snakes are not poisonous.
(b) Some snakes are poisonous.
(c) Some snakes are not poisonous.

Solution. The domain of discourse U is the collection of all snakes. Let $p(x)$ be the propositional predicate which asserts that x is poisonous, (where the variable x ranges over U). The statement "All snakes are poisonous" is then translated into $(\forall x)\ (p(x))$. According to Q.N., $\sim[(\forall x)\ (p(x))]$ is equivalent to $(\exists x)\ (\sim p(x))$, which represents "Some snakes are not poisonous."

EXERCISE 1.6

1. Translate the statement from elementary algebra "The equation $x^2 - 3x + 2 = 0$ has solutions" into logical language using a quantifier. What is the domain of discourse here?
2. Translate the identity "$(x + a)^2 = x^2 + 2ax + a^2$" from elementary algebra into logical language using a quantifier. What is the domain of discourse here?
3. A generalized De Morgan's Law may be stated as:
   (a) $\sim(p_1 \wedge p_2 \wedge \cdots \wedge p_n) \equiv \sim p_1 \vee \sim p_2 \vee \cdots \vee \sim p_n$
   (b) $\sim(p_1 \vee p_2 \vee \cdots \vee p_n) \equiv \sim p_1 \wedge \sim p_2 \wedge \cdots \wedge \sim p_n$
   Translate these into logical language using suitable quantifiers with a suitable domain of discourse. What is the domain of discourse here?
4. Find the equivalent statement of the negation for each of the following by using Q.N.
   (a) All snakes are reptiles.
   (b) Some horses are gentle.
   (c) Some mathematicians are not sociable.
   (d) All female students are either attractive or smart.
   (e) No baby is not cute.
5. Find the domain of discourse for each of the five statements in problem 4.
6. Derive
$$\sim[(\exists x)\ (p(x))] \equiv (\forall x)\ (\sim p(x))$$
   from
$$\sim[(\forall x)\ (q(x))] \equiv (\exists x)\ (\sim q(x)).$$
7. Derive
$$\sim[(\forall x)\ (p(x))] \equiv (\exists x)\ (\sim p(x))$$
   from
$$\sim[(\exists x)\ (q(x))] \equiv (\forall x)\ (\sim q(x)).$$

8. Prove that $\sim[(\forall x)(\sim q(x))] \equiv (\exists x)(q(x))$ and $\sim[(\exists x)(\sim q(x))] \equiv (\forall x)(q(x))$. (Hint: Use Q.N.]

7. PROOF OF VALIDITY

One of the most important tasks of a logician is the testing of *arguments*. An argument is the assertion that a statement, called the *conclusion*, follows from other statements, called the *hypotheses* or *premises*. An argument is considered to be *valid* if the conjunction of the hypotheses implies the conclusion. For examples, the following is an argument in which the first four statments are hypotheses, and the last statement is the conclusion.

If he studies medicine, then he prepares to earn a good living.
If he studies the arts, then he prepares to live a good life.
If he prepares to earn a good living or he prepares to live a good life, then his college tuition is not wasted.
His college tuition is wasted.
Therefore, he studies neither medicine nor the arts.

It may be symbolized as:

H1. $M \to E$
H2. $A \to L$
H3. $(E \vee L) \to \sim W$
H4. W

C. $\therefore \sim M \wedge \sim A$

To establish the validity of this argument by means of a truth table would require a table with 32 $(=2^5)$ rows. But we can prove this argument valid by deducing the conclusion from the hypotheses in a few steps using the rules of inference.

From the hypotheses H3 and H4, $(E \vee L) \to \sim W$ and W, we infer $\sim(E \vee L)$, or equivalently $\sim E \wedge \sim L$, by Modus Tollens and De Morgan's Law. From $\sim E \wedge \sim L$ we validly infer $\sim E$ (and also $\sim L$) by the Law of Simplification. From H1, $M \to E$, and $\sim E$ we validly obtain $\sim M$. Similarly, from H2, $A \to L$, and $\sim L$, we validly infer $\sim A$. Finally, the conjunction of $\sim M$ and $\sim A$ gives the conclusion $\sim M \wedge \sim A$. In this proof, the rules of inferences Modus Tollens (M.T.), De Morgan's Law (De M.), and Laws of Simplification (Simp.) are used.

A formal and more concise way of expressing this proof of validity is to list the hypotheses and the statements deduced from them on one side, with the justification of each step written beside it. In each step the "justification" indicates the preceeding statements from which, and the rules of inference by which, the statement given in that step was obtained. For easy reference, it is convenient to number the hypotheses and the statements deduced from them and to put the conclusion to the right of the last premise, separated from it by a slash / which indicates that all the statements above it are hypotheses. The formal proof of validity for the above argument may thus be written as

1.	$M \to E$	(Hyp.)
2.	$A \to L$	(Hyp.)
3.	$(E \vee L) \to \sim W$	(Hyp.)
4.	$W \,/ \therefore \sim M \wedge \sim A$	(Hyp. / Concl.)
5.	$\sim(E \vee L)$	3, 4, M.T.
6.	$\sim E \wedge \sim L$	5, De M.
7.	$\sim E$	6, Simp.
8.	$\sim L$	6, Simp.
9.	$\sim M$	1, 7, M.T.
10.	$\sim A$	2, 8, M.T.
11.	$\sim M \wedge \sim A$	9, 10, Conj.

In step 11 above, we have used the following clearly valid argument, called *Conjunction* (abbreviation: Conj.):

H 1.	p
H 2.	q
C.	$\therefore p \wedge q$

A *formal proof of validity* for a given argument is a sequence of statements each of which is either a premise of the argument or follows from preceeding statments by a known valid argument, ending with the conclusion of the argument.

EXAMPLE 10. Construct a formal proof of validity for the following argument, using the suggested symbols:

Either Winston is elected president of the board or both Halbert and Luke are elected vice presidents of the board. If either Winston is elected president or Halbert is elected vice president of the board, then David will file a protest. Therefore, either Winston is elected president of the board or David files a protest. (W, H, L, D)

Proof.

1.	$W \vee (H \wedge L)$	
2.	$W \vee H \to D \,/ \therefore W \vee D$	
3.	$(W \vee H) \wedge (W \vee L)$	1, Dist.
4.	$W \vee H$	3, Simp.
5.	D	2, 4, M.P.
6.	$D \vee W$	5, Add.
7.	$W \vee D$	6, Com.

There is another method of proof called *indirect proof*, or the method of proof by *reductio ad absurdum*. An indirect proof of validity for a given argument is done by including, as an additional premise, the negation of its conclusion, and then deriving a contradiction; as soon as a contradiction is obtained, the proof is complete.

EXAMPLE 11. Give an indirect proof of validity for the following argument:

$$p \vee q \rightarrow r$$
$$s \rightarrow p \wedge u$$
$$q \vee s \;/\; \therefore r$$

Proof.

1.	$p \vee q \rightarrow r$	
2.	$s \rightarrow p \wedge u$	
3.	$q \vee s \;/\; \therefore r$	
4.	$\sim r$	I.P. (Indirect Proof)
5.	$\sim (p \vee q)$	1, 4, M.T.
6.	$\sim p \wedge \sim q$	5, De M.
7.	$\sim p$	6, Simp.
8.	$\sim q$	6, Simp.
9.	s	3, 8, D.S.
10.	$p \wedge u$	2, 9, M.P.
11.	p	10, Simp.
12.	$p \wedge \sim p$	7, 11, Conj.

The statement $p \wedge \sim p$ in step 12 is a contradiction; therefore the indirect proof of validity is complete.

In contrast to an "indirect proof" the formal proof of validity introduced earlier may be called a "direct proof." In a mathematical proof, a direct proof or an indirect proof may be used. The choice of the method of proof for a given mathematical argument depends on taste and convenience.

EXERCISE 1.7

Each of problems 1 through 4 is a formal proof of validity for the indicated argument. State the justification for each line that is not a premise:

1. $p \vee q \rightarrow r \wedge s$
   $\sim r \;/\; \therefore \sim q$
   $\sim r \vee \sim s$
   $\sim (r \wedge s)$
   $\sim (p \vee q)$
   $\sim p \wedge \sim q$
   $\sim q$

2. $p \wedge q \rightarrow r$
$(p \rightarrow r) \rightarrow s$
$\sim q \vee u \ / \ \therefore \ q \rightarrow s \wedge u$
$q \wedge p \rightarrow r$
$q \rightarrow (p \rightarrow r)$
$q \rightarrow s$
$\sim q \vee s$
$(\sim q \vee s) \wedge (\sim q \vee u)$
$\sim q \vee (s \wedge u)$
$q \rightarrow s \wedge u$

3. $p \wedge (q \vee s)$
$p \rightarrow (s \rightarrow u)$
$p \rightarrow (q \rightarrow r)$
$\sim r \ / \ \therefore \ u$
$p \wedge s \rightarrow u$
$p \wedge q \rightarrow r$
$(p \wedge q) \vee (p \wedge s)$
$(p \wedge q \rightarrow r) \wedge (p \wedge s \rightarrow u)$
$(p \wedge q) \vee (p \wedge s) \rightarrow r \vee u$
$r \vee u$
u

4. $\sim p \wedge s \wedge \sim(\sim p \wedge q)$
$p \vee q \vee (r \wedge s) \ / \ \therefore \ \sim p \wedge r$
$\sim p \wedge s$
$\sim p$
$p \vee [q \vee (r \wedge s)]$
$q \vee (r \wedge s)$
$(q \vee r) \wedge (q \vee s)$
$q \vee r$
$\sim p \wedge (q \vee r)$
$(\sim p \wedge q) \vee (\sim p \wedge r)$
$(\sim p \wedge s) \wedge \sim(\sim p \wedge q)$
$\sim(\sim p \wedge q)$
$\sim p \wedge r$

For each of the following arguments give either a direct proof or an indirect proof of validity.

5. $A \vee (B \wedge C)$
$B \rightarrow D$
$C \rightarrow E$
$D \wedge E \rightarrow F$
$\sim A \ / \ \therefore \ F$

6. $B \vee (C \rightarrow E)$
$B \rightarrow D$
$\sim D \rightarrow (E \rightarrow A)$
$\sim D \ / \ \therefore C \rightarrow A$

7. $(A \vee B) \rightarrow (A \rightarrow D \wedge E)$
$A \wedge C \ / \ \therefore E \vee F$

8. $A \vee B$
$\sim B \vee C \ / \ \therefore A \vee C$

9. $B \vee C \rightarrow B \wedge A$
$\sim B \ / \ \therefore \ \sim C$

10. $A \wedge B \rightarrow C$
$(A \rightarrow C) \rightarrow D$
$\sim B \vee E \ / \ \therefore B \rightarrow D \wedge E$

In the proofs of the following arguments, use the suggested symbols.

11. If the population increases rapidly and production remains constant, then prices rise. If prices rise then the government will control prices. If I am rich then I do not care about increase in prices. It is not true that I

am not rich. Either the government does not control prices or I do care about increase in prices. Therefore, it is not true that the population increases rapidly and production remains constant. (P: The population increases rapidly. C: production remains constant. R: prices rise. G: The government will control prices. H: I am rich. I: I care about increase in prices.)

12. Dean praises me only if I can be proud of myself. Either I do well in classes or I cannot be proud of myself. If I do my best in sports, then I cannot do well in classes. Therefore, if Dean praises me, then I do not do my best in sports. (D: Dean praises me. P: I can be proud of myself. C: I do well in classes. S: I do my best in sports.)
13. If Winston or Halbert wins then Luke and Susan cry. Susan does not cry. Therefore, Halbert did not win. (W: Winston wins. H: Halbert wins. L: Luke cries. S: Susan cries.)
14. If I enroll in this course and study hard then I make good grades. If I make good grades then I am happy. I am not happy. Therefore, either I did not enroll in this course or I did not study hard. (E: I enroll in this course. S: I study hard. G: I make good grades. H: I am happy.)
15. [Adapted from Lewis Carroll.]
If he goes to a party, he does not fail to brush his hair. To look fascinating it is necessary to be tidy. If he is an opium eater, then he has no self-command. If he brushes his hair, he looks fascinating. He wears white kid gloves only if he goes to a paty. Having no self-command is sufficient to make one look untidy. Therefore, opium eaters do not wear white kid gloves. (P: He goes to a party. B: He does not brush his hair. F: He looks fascinating. T: He looks tidy. O: He is an opium eater. S: He has self-command. G: He wears white kid gloves.)

8. MATHEMATICAL INDUCTION

Another method of proof that is very useful in proving the validity of a mathematical statment $P(n)$ involving the natural number n is the following principle of *mathematical induction.*

Mathematical Induction. If $P(n)$ is a statement involving the natural number n such that

(1) $P(1)$ is true, and
(2) $P(k) \Rightarrow P(k+1)$ for any arbitrary natural number k,

then $P(n)$ is true for every natural number n.

The above principle is a consequence of one of Peano's Axioms for the natural numbers, which are included in the Appendix for reference.

In order to apply the principle of mathematical induction to prove a theorem, the theorem must be capable of being broken down into cases, one case for each natural number. Then, we must verify both conditions (1) and (2). The verification of (1), that usually is easy, assures us that the theorem is true for as least the case $n = 1$. To verify condition (2), we must prove an auxiliary theorem whose premise is "$P(k)$ is true" and whose conclusion is "$P(k + 1)$ is true." The premise "$P(k)$ is true" is called the *induction hypothesis*.

EXAMPLE 12. Prove by mathematical induction that

$$1 + 2 + 3 + \cdots + n = \frac{n(n + 1)}{2} \quad \text{for every natural number } n$$

Proof. Here $P(n)$ represents the statement

$$\text{“}1 + 2 + 3 + \cdots + n = \frac{n(n + 1)}{2}\text{”}$$

In particular, $P(1)$ represents "$1 = (1 \cdot 2) / 2$" which obviously is a true statement. Therefore, condition (1) for mathematical induction is satisfied.

To prove that condition (2) is satisfied, we assume that $P(k)$, which is "$1 + 2 + 3 + \cdots + k = k(k + 1) / 2$," is true. Then add $k + 1$ to both sides of this equality. We have

$$\begin{aligned} 1 + 2 + 3 + \cdots + k + (k + 1) &= \frac{k(k + 1)}{2} + (k + 1) \\ &= \frac{k(k + 1)}{2} + \frac{2(k + 1)}{2} \\ &= \frac{(k + 2)(k + 1)}{2} \\ &= \frac{(k + 1)(k + 2)}{2} \end{aligned}$$

which shows that $P(k + 1)$ is true. We have now shown that the conditions (1) and (2) of mathematical induction are satisfied. Therefore, by the principle of mathematical induction, $1 + 2 + 3 + \cdots + n = n(n + 1)/2$ is true for every natural number n.

The idea of mathematical induction may be used in making definitions involving natural numbers. For example, the definition of *powers* of an unknown number x may be defined by:

$$x^1 = x$$

$$x^{n+1} = x^n \cdot x, \qquad \text{for any natural number } n$$

The above two equations indicate that $x^1 = x, x^2 = x \cdot x, x^3 = x^2 \cdot x, \ldots$ and so forth. As another application we give the following inductive definition of the symbol $C(n, r)$.

Definition 7. Let n be a natural number and r an integer. The symbol $C(n, r)$ is defined by

$$C(0, 0) = 1, \quad C(0, r) = 0 \quad \text{for each } r \neq 0, \text{ and}$$
$$C(n, r) = C(n - 1, r) + C(n - 1, r - 1)$$

Theorem 8. If n and r are integers such that $0 \leq r \leq n$, then

$$C(n, r) = \frac{n!}{r!(n - r)!}$$

where $n!$ denotes the product $n \cdot (n - 1) \cdots 3 \cdot 2 \cdot 1$ of the first n consecutive natural numbers if $n > 0$ and $0! = 1$ by convention.

Proof. Exercise.

Theorem 9. (*The Binomial Theorem*). If x and y are two variables and n is a natural number, then

$$(x + y)^n = C(n, 0)x^n + C(n, 1)x^{n-1}y + \cdots + C(n, r)x^{n-r}y^r + \cdots + C(n, n)y^n$$

Proof. We shall prove the validity of this theorem by mathematical induction. First, the theorem is clearly true for $n = 1$. To complete the proof, we assume the validity of the theorem for $n = k$; that is, we assume that

$$(x + y)^k = C(k, 0)x^k + C(k, 1)x^{k-1}y + \cdots + C(k, r)x^{k-r}y^r + \cdots + C(k, k)y^k$$

Then, by multiplying both sides of the above equality by $(x + y)$, we have

$$\begin{aligned}(x + y)^{k+1} &= (x + y)\,[x^k + C(k, 1)x^{k-1}y + \cdots + C(k, r)x^{k-r}y^r + \cdots + y^k] \\ &= x^{k+1} + [C(k, 0) + C(k, 1)]x^k y + \cdots \\ &\quad + [C(k, r - 1) + C(k, r)]x^{(k+1)-r}y^r + \cdots + y^{k+1} \\ &= C(k + 1, 0)x^{k+1} + C(k + 1, 1)x^k y + \cdots \\ &\quad + C(k + 1, r)x^{k+1-r}y^r + \cdots + C(k + 1, k + 1)y^{k+1}\end{aligned}$$

which shows that the theorem is valid for $n = k + 1$ if it is valid for $n = k$. Thus, by mathematical induction, the binomial theorem is true for all natural numbers n.

The numbers $C(n, r)$ in the binomial theorem are called *binomial coefficients.*

EXERCISE 1.8

1. Prove Theorem 8 by mathematical induction.
2. Show that $C(n, 0) = 1 = C(n, n)$ for all natural numbers n.
3. Prove by mathematical induction that for all natural numbers n, $C(n, 0) + C(n, 1) + \cdots + C(n, n) = 2^n$.
4. Prove by mathematical induction that for all natural numbers n, $1 + 2 + 3 + \cdots + n = n(n + 1)/2$.
5. Prove by mathematical induction that for all natural numbers n.

$$1 \cdot 2 + 2 \cdot 3 + \cdots + r \cdot (r + 1) + \cdots + n(n + 1) = n(n + 1)(n + 2)/3.$$

6. Prove by mathematical induction that for all natural numbers n,

$$1^2 + 2^2 + 3^2 + \cdots + n^2 = n(n + 1)(2n + 1)/6.$$

7. Prove that for all natural numbers n,

$$1^3 + 2^3 + 3^3 + \cdots + n^3 = n^2(n + 1)^2/4.$$

8. Prove that for all natural numbers n,
$$(1 + 2 + 3 + \cdots + n)^2 = 1^3 + 2^3 + 3^3 + \cdots + n^3.$$

9. Prove that for all natural numbers n,

$$1 + 3 + 5 + \cdots + (2n - 1) = n^2.$$

10. Prove that for all natural numbers n,

$$\frac{1}{1 \cdot 2} + \frac{1}{2 \cdot 3} + \frac{1}{3 \cdot 4} + \cdots + \frac{1}{n(n + 1)} = \frac{n}{n + 1}$$

11. Prove the following Generalized De Morgan's Laws.

(a) $\sim(p_1 \wedge p_2 \wedge \cdots \wedge p_n) \Leftrightarrow \sim p_1 \vee \sim p_2 \vee \cdots \vee \sim p_n$
(b) $\sim(p_1 \vee p_2 \vee \cdots \vee p_n) \Leftrightarrow \sim p_1 \wedge \sim p_2 \wedge \cdots \wedge \sim p_n$

12. Prove the following Generalized Distributive Laws.

(a) $p \wedge (q_1 \vee q_2 \vee \cdots \vee q_n) \Leftrightarrow (p \wedge q_1) \vee (p \wedge q_2) \vee \cdots \vee (p \wedge q_n)$

(b) $p \vee (q_1 \wedge q_2 \wedge \cdots \wedge q_n) \Leftrightarrow (p \vee q_1) \wedge (p \vee q_2) \wedge \cdots \wedge (p \vee q_n)$

13. Prove that if k is any fixed natural number, then for all natural numbers n,

$$1 \cdot 2 \cdots k + 2 \cdot 3 \cdots (k+1) + \cdots + n(n+1) \cdots (n+k-1)$$
$$= \frac{1}{k+1} n(n+1)(n+2) \cdots (n+k).$$

Chapter Two

The Concept of Sets

In this chapter, we introduce the concept of sets, subsets, and set operations (union, intersection, and complementation) together with the fundamental rules governing these operations. These are developed in parallel with Chapter 1 on logic. Indexed families of sets are discussed. The chapter ends with the Russell Paradox and a historical remark.

1. SETS AND SUBSETS

"What is a set?" is a very difficult question to answer.[1] In this elementary book, we shall not go into any complicated axiomatic approach to Set Theory, but shall content ourself to accept the following: a *set* is any collection into a whole of definite, distinguishable objects, called *elements*, of our intuition or thought. This intuitive definition of a set was first given by Georg Cantor (1845-1918), who originated the theory of sets in 1895. Examples:

(a) The set of all chairs in this classroom.
(b) The set of all students in this university.
(c) The set of letters a, b, c, and d.
(d) The set of rules in our dormitories.
(e) The set of all rational numbers whose square is 2.
(f) The set of all natural numbers.
(g) The set of all real numbers between 0 and 1.

---

[1] Students will realize the difficulty when they come to Sections 7 and 8.

A set which contains only finitely many elements is called a *finite set*; an *infinite set* is one which is not a finite set. Examples (a) to (e) above are all finite sets, and Examples (f) and (g) are infinite sets.

Sets are frequently designate by enclosing symbols representing their elements in braces when it is possible to do so. Thus, the set in Example (c) is $\{a, b, c, d\}$ and the set in Example (f) may be denoted by $\{1, 2, 3, \ldots\}$. The set described in Example (e) has no elements; such a set is called the *empty set*, which will be denoted by the symbol $\varnothing$.

We shall also use capitol letters to denote sets, and lower-case letters to denote elements. If a is an element of the set A, we write $a \in A$ (read: "a is an element of A" or "a belongs to A,") whereas $b \notin A$, means that b is not an element of A.

Definition 1. Two set A and B are said to be *equal* or identical, in symbols: $A = B$, provided that they contain the same elements. That is, $A = B$ means $(\forall x)\ [(x \in A) \leftrightarrow (x \in B)]$.

The order of appearance of the elements of a set is of no consequence. Thus, the set $\{a, b, c\}$ is the same as $\{b, c, a\}$ or $\{c, b, a\}$, etc. Furthermore, since elements in a set are distinct, $\{a, a, b\}$, for example, is not a proper notation of a set and should be replaced by $\{a, b\}$. If a is an element of a set, a and $\{a\}$ are to be considered different, that is, $a \neq \{a\}$. For, $\{a\}$ denotes the *set* consisting of the single element a alone, whereas a is just the *element* in the set $\{a\}$.

Definition 2. Let A and B be sets. If every element of A is an element of B, then A is called a *subset* of B, in symbols: $A \subseteq B$ or $B \supseteq A$. If A is a subset of B, then B is called a *superset* of A.

Thus, logically speaking,

$$A \subseteq B \equiv (\forall x)\ [(x \in A) \rightarrow (x \in B)]$$

Obviously, every set is a subset (and a superset) of itself. When $A \subseteq B$ and $A \neq B$, we write $A \subset B$, or $B \supset A$, which reads: A is a *proper* subset of B, or B is a *proper* superset of A. In other words, A is a proper subset of B provided that every element of A is an element of B, and there exists an element of B which is not an element of A. If A is not a subset of B, we write $A \nsubseteq B$.

Theorem 1. The empty set $\varnothing$ is a subset of every set.

Proof. Let A be any set. We are to prove that the conditional statement

$$(x \in \varnothing) \rightarrow (x \in A)$$

is true for every x. Since the empty set $\varnothing$ has no elements, the statement "$x \in A$" may be true or may be false. In either case, the conditional statement $(x \in \varnothing) \rightarrow (x \in A)$ is true according to the truth table for the conditional (case 3 and 4 of Table 5, Chapter 1). Thus, $\varnothing \subseteq A$ for any set A.

Theorem 2. If $A \subseteq B$ and $B \subseteq C$, then $A \subseteq C$.

Proof. We are to show that $(x \in A) \Rightarrow (x \in C)$:

$$(x \in A) \Rightarrow (x \in B), \quad \text{because} \quad A \subseteq B$$

$$\Rightarrow (x \in C), \quad \text{because} \quad B \subseteq C$$

Hence, by the Transitive Law (Theorem 4(c) of Chapter 1), we have

$$(x \in A) \Rightarrow (x \in C)$$

Thus, we have proved that $A \subseteq C$.

EXERCISE 2.1

1. Show that the set of letters needed to spell "cataract" and the set of letters needed to spell "tract" are equal.
2. Decide, among the following sets, which are subsets of which.
   (a) $A = \{\text{all real numbers satisfying } x^2 - 8x + 12 = 0\}$
   (b) $B = \{2, 4, 6\}$
   (c) $C = \{2, 4, 6, 8, \ldots\}$
   (d) $D = \{6\}$
3. List all the subsets of the set $\{-1, 0, 1\}$.
4. Prove that $[(A \subseteq B) \wedge (B \subseteq A)] \Leftrightarrow (A = B)$. [*Remark*: Frequently in mathematics the best way to show that $A = B$ is to show that $A \subseteq B$ and $B \subseteq A$.]
5. Prove that $(A \subseteq \varnothing) \Rightarrow (A = \varnothing)$.
6. Prove that
   (a) $[(A \subset B) \wedge (B \subseteq C)] \Rightarrow (A \subset C)$
   (b) $[(A \subseteq B) \wedge (B \subset C)] \Rightarrow (A \subset C)$
7. Give an example of a set whose elements are themselves sets.
8. In each of the following, determine whether the statement is true or false. If it is true, prove it. If it is false, disprove it by an example (such an example which disproves a statement is called a counterexample).
   (a) If $x \in A$ and $A \in B$, then $x \in B$.
   (b) If $A \subseteq B$ and $B \in C$, then $A \in C$.
   (c) If $A \nsubseteq B$ and $B \subseteq C$, then $A \nsubseteq C$.
   (d) If $A \nsubseteq B$ and $B \nsubseteq C$, then $A \nsubseteq C$.

(e) If $x \in A$ and $A \nsubseteq B$, then $x \notin B$.
(f) If $A \subseteq B$ and $x \notin B$, then $x \notin A$.

9. Determine whether the following statements are true or false:
(a) $x \in \{\{x\}, \{x, y\}\}$
(b) $\{x\} \subseteq \{\{x\}, \{x, y\}\}$
(c) $\{1, x, 2\} \subseteq \{1, 2, x\}$
(d) $\{a, b\} \subseteq \{b, a\}$
(e) $\{x\} \in \{x\}$.

10. Determine whether the following statements are true or false:
(a) $\varnothing = \{\varnothing\}$
(b) $\varnothing \in \{\varnothing\}$
(c) $\{\varnothing\} \in \varnothing$
(d) $\{\varnothing, \{\varnothing\}\} \in \{\varnothing, \{\varnothing\}, \{\varnothing, \varnothing\}\}$.

11. Given a set with n elements, prove that there are exactly $C(n, r)$ subsets with r elements.

2. SPECIFICATION OF SETS

One way of making a new set out of a given set is to specify those elements of the given set that satisfy a particular property. For example, let A be the set of all students in this university. The statement "x is female" is true for some elements x of A and false for others. We shall use the notation

$$\{x \in A \mid x \text{ is female}\}$$

to specify the set of all female students in this university. Similarly,

$$\{x \in A \mid x \text{ is not female}\}$$

specifies the set of all male students in this university.

As a rule, to every set A and to every statement $p(x)$ about $x \in A$, there exists a set $\{x \in A \mid p(x)\}$ whose elements are precisely those elements x of A for which the statement $p(x)$ is true. In an axiomatic approach to set theory, this rule is usually postulated as an axiom, called the *Axiom of Specification*. The symbol $\{x \in A \mid p(x)\}$ reads: the set of all x in A such that $p(x)$ is true. The notation of the form $\{x \in A \mid p(x)\}$ which describes a set is called the *set builder notation.*

EXAMPLE 1. Let $\mathbf{R}$ denote the set of all real numbers. Then

(a) $\{x \in \mathbf{R} \mid x = x+1\}$ is the empty set.
(b) $\{x \in \mathbf{R} \mid 2x^2 - 5x - 3 = 0\}$ is the set $\{-\frac{1}{2}, 3\}$.
(c) $\{x \in \mathbf{R} \mid x^2 + 1 = 0\}$ is the empty set.

Because of their frequent appearance throughout the remainder of this book and in other topics of mathematics, the following special symbols will be reserved for the sets described:

$$\begin{aligned}
\mathbf{R} &= \{x \mid x \text{ is a real number}\}\\
\mathbf{Q} &= \{x \mid x \text{ is a rational number}\}\\
\mathbf{Z} &= \{x \mid x \text{ is an integer}\}\\
\mathbf{N} &= \{n \mid n \text{ is a natural number}\}\\
\mathbf{I} &= \{x \in \mathbf{R} \mid 0 \le x \le 1\}\\
\mathbf{R}_{+} &= \{x \in \mathbf{R} \mid x > 0\}
\end{aligned}$$

It should be noticed that $\mathbf{N} \subset \mathbf{Z} \subset \mathbf{Q} \subset \mathbf{R}$ and $\mathbf{N} \subset \mathbf{R}_{+} \subset \mathbf{R}$.

It is quite possible that elements of a set may themselves be sets. For example, the set of all subsets of a given set A has sets as its elements. This set is called the *power set* [2] of A and is denoted by $\mathscr{P}(A)$.

EXAMPLE 2. $\mathscr{P}(\{a\}) = \{\varnothing, \{a\}\}$, $\mathscr{P}(\varnothing) = \{\varnothing\}$, and $\mathscr{P}(\{a, b\}) = \{\varnothing, \{a\}, \{b\}, \{a,b\}\}$.

The name "power set" is motivated by the following theorem.

Theorem 3. If A consists of n elements, then its power set $\mathscr{P}(A)$ contains exactly 2^n elements.

Proof. The theorem is clearly true for $A = \varnothing$. For a nonempty set A, we let $A = \{a_1, a_2, a_3, \ldots, a_n\}$. Given an element a_k of A, each subset of A has two possibilities: it either contains a_k or it does not. Therefore, the problem of finding the number of subsets of A may be considered as the problem of filling a list of n blank spaces □ □ □ · · · □ at random with the numbers 0 and 1, one number in each space. Each filling of the n blanks determines a subset X of A in the following manner: $a_k \in X$ if and only if 1 appears in the kth space. Since there are exactly 2^n different such fillings, there are exactly 2^n subsets of A.

It may be interesting to know the following alternative proof of Theorem 3.

Alternative Proof. First the empty set $\varnothing$ belongs to $\mathscr{P}(A)$. Next, each element $x \in A$ forms a subset $\{x\}$ belonging to $\mathscr{P}(A)$. Observe that the number of

[2] In an axiomatic set theory, the existence of the power set is not taken for granted. Since the existence of a power set does not follow from the axiom of specification, a new axiom is needed; this axiom is usually called the *Axiom of Power Sets* and may be stated: *For each set there exists a set of sets that consists of all the subsets of the given set.*

these singleton subsets is $C(n, 1)$. Continuing, there are exactly $C(n, 2)$ subsets of A containing exactly two elements of A.[3] Finally, there is exactly $C(n, n) = 1$ subset of A containing n elements of A, namely the set A itself. Counting the empty set, the total number of subsets of A is equal to $C(n, 0) + C(n, 1) + \cdots + C(n, n)$. Then using the binomial expansion for $(1+1)^n$, we have

$$(1+1)^n = C(n, 0) + C(n, 1) + \cdots + C(n, n)$$

Thus, the number of elements in $\mathscr{P}(A)$ is $(1+1)^n = 2^n$.

EXERCISE 2.2

1. Display within braces the elements of each of the following sets.

$$A = \{x \in \mathbf{N} \mid x < 5\}$$

$$B = \{x \in \mathbf{Z} \mid x^2 \leq 25\}$$

$$C = \{x \in \mathbf{Q} \mid 10x^2 + 3x - 1 = 0\}$$

$$D = \{x \in \mathbf{R} \mid x^3 + 1 = 0\}$$

$$E = \{x \in \mathbf{R}_+ \mid 4x^2 - 4x - 1 = 0\}$$

2. Display within braces the elements of each of the following sets:
$A = \{x \in N \mid 5 < x < 3\}$
$B = \{x \in Z \mid 25 \leq x^2 < 36\}$
$C = \{x \in Q \mid x^3 - 6x^2 + 11x - 6 = 0\}$
$D = \{x \in \text{Europe} \mid x \text{ is a capital city}\}$.
3. Denote each of the following sets by the set builder notation:
$A = \{1, 4, 9, 16, 25, \ldots\}$
$B = \{2, 4, 6, 8, 10, \ldots\}$
$C = \{0, \frac{1}{2}, \frac{2}{3}, \frac{3}{4}, \ldots\}$
$D = \{2, 9, 28, 65, \ldots\}$
$E = \{a, e, i, o, u\}$
4. Denote each of the following sets by the set builder notation.

$$A = \{1, 2, 3\}$$

$$B = \{-1, -\tfrac{2}{3}, -\tfrac{1}{3}, 0\}$$

[3] See problem 11, Exercise 2.1.

$$C = \{1, 3, 5, 7, 9, \ldots\}$$

$$D = \{1-\sqrt{3}, 1+\sqrt{3}\}$$

5. What are the elements of the power set of the set $\{x, \{y, z\}\}$? How many elements does this power set contain?
6. Is the power set of the empty set empty? Why?
7. How many elements in the power set of the power set of the empty set?
8. Construct the set $\mathscr{P}(\mathscr{P}(\mathscr{P}(\varnothing)))$. Does this set contain more elements than the set $\mathscr{P}(\mathscr{P}(\varnothing))$?
9. Is $\mathscr{P}(A) \cup \mathscr{P}(B) = \mathscr{P}(A \cup B)$ for any sets A and B? Why?
10. Is $\mathscr{P}(A) \cap \mathscr{P}(B) = \mathscr{P}(A \cap B)$ for any sets A and B? Why?
11. Let B be a subset of A, and let $\mathscr{P}(A: B) = \{X \in \mathscr{P}(A) \mid X \supseteq B\}$.
    (a) Let $B = \{a, b\}$ and $A = \{a, b, c, d, e\}$. List all the members of the set $\mathscr{P}(A: B)$; how many are there?
    (b) Show that $\mathscr{P}(A:\varnothing) = \mathscr{P}(A)$.
12. Let A be a set with n elements and B a subset with m elements, $n \geq m$.
    (a) Find the number of elements in the set $\mathscr{P}(A: B)$.
    (b) Deduce Theorem 3 from (a) by setting $B = \varnothing$.

3. UNIONS AND INTERSECTIONS

In arithmetic we can add, multiply, or subtract any two numbers. In set theory, there are three operations—union, intersection, and complementation—analogous respectively to the addition, multiplication, and subtraction of numbers.

Definition 3. The *union* of any two sets A and B, denoted by $A \cup B$, is the set of all elements x such that x belongs to at least one of the two sets A and B. That is, $x \in A \cup B$ if and only if $x \in A \vee x \in B$.

Definition 4. The *intersection* of any two sets A and B, denoted by $A \cap B$, is the set of all elements x which belong to both A and B. In symbols, $A \cap B = \{x \mid (x \in A) \wedge (x \in B)\}$, or $\{x \in A \mid x \in B\}$. If $A \cap B = \varnothing$, then A and B are said to be *disjoint*.

For example, if $A = \{1, 2, 3, 4\}$ and $B = \{3, 4, 5\}$ then $A \cup B = \{1, 2, 3, 4, 5\}$ and $A \cap B = \{3, 4\}$; if **Im** denote the set of imaginary numbers, then the sets **Im** and **R** are disjoint.

EXAMPLE 3. In the following, the sets **I**, **N**, **Z**, . . . are as defined in the last section.

(a) $\mathbf{I} \cap \mathbf{Z} = \{0, 1\}$ and $\mathbf{N} \cap \mathbf{I} = \{1\}$.
(b) $\mathbf{Z} \cup \mathbf{Q} = \mathbf{Q}$ and $\mathbf{Z} \cap \mathbf{Q} = \mathbf{Z}$.
(c) $\mathbf{I} \cup \mathbf{I} = \mathbf{I}$ and $\mathbf{I} \cap \mathbf{I} = \mathbf{I}$.

Theorem 4. Let X be a set and let A, B, and C be subsets of X. Then we have

(a) The unities:
$$A \cup \varnothing = A$$
$$A \cap X = A$$
(b) The idempotency laws:
$$A \cup A = A$$
$$A \cap A = A$$
(c) The commutative laws:
$$A \cup B = B \cup A$$
$$A \cap B = B \cap A$$
(d) The associative laws:
$$A \cup (B \cup C) = (A \cup B) \cup C$$
$$A \cap (B \cap C) = (A \cap B) \cap C$$
(e) The distributive laws:
$$A \cap (B \cup C) = (A \cap B) \cup (A \cap C)$$
$$A \cup (B \cap C) = (A \cup B) \cap (A \cup C)$$

Proof. We leave the proofs of parts (a), (b), and (c) to the reader as exercises.

(d). According to Definition 3, an element
$$x \in A \cup (B \cup C) \Leftrightarrow x \in A \vee (x \in B \cup C)$$
and
$$x \in B \cup C \Leftrightarrow x \in B \vee x \in C$$
so
$$x \in A \cup (B \cup C) \Leftrightarrow x \in A \vee (x \in B \vee x \in C)$$

By the Associative Law (for the disjunction), $(x \in A) \vee (x \in B \vee x \in C)$ is equivalent to $(x \in A \vee x \in B) \vee (x \in C)$. The last statement, by Definition 3, is equivalent to $(x \in A \cup B) \vee (x \in C)$, and hence to $x \in (A \cup B) \cup C$. Thus, we have
$$x \in A \cup (B \cup C) \Leftrightarrow x \in (A \cup B) \cup C$$

By Definition 1, $A \cup (B \cup C) = (A \cup B) \cup C$

The above proof may be condensed into a neat display of essential logical steps, with the justification of each step written on the right for easy reference:

$x \in A \cup (B \cup C) \Leftrightarrow (x \in A) \vee (x \in B \cup C)$	Def. $\cup$
$\Leftrightarrow (x \in A) \vee [(x \in B) \vee (x \in C)]$	Def. $\cup$
$\Leftrightarrow [(x \in A) \vee (x \in B)] \vee (x \in C)$	Assoc. for $\vee$
$\Leftrightarrow (x \in A \cup B) \vee (x \in C)$	Def. $\cup$
$\Leftrightarrow x \in (A \cup B) \cup C$	Def. $\cup$

Hence, by Definition 1, we have proved that $A \cup (B \cup C) = (A \cup B) \cup C$.

The students should try to appreciate this kind of orderly precise proof by logic.

We leave the proof of $A \cap (B \cap C) = (A \cap B) \cap C$ to the reader as an exercise.

(e) Again only the first half of (e) is proved; the other half is left to the reader as an exercise.

$x \in [A \cap (B \cup C)] \Leftrightarrow [(x \in A) \wedge (x \in B \cup C)]$	Def. $\cap$
$\Leftrightarrow (x \in A) \wedge [(x \in B) \vee (x \in C)]$	Def. $\cup$
$\Leftrightarrow [(x \in A) \wedge (x \in B)] \vee [(x \in A) \wedge (x \in C)]$	Dist. Law of logic (Ch. 1)
$\Leftrightarrow [(x \in A \cap B) \vee (x \in A \cap C)]$	Def. $\cap$
$\Leftrightarrow x \in [(A \cap B) \cup (A \cap C)]$	Def. $\cup$

Hence, by Definition 1, $A \cap (B \cup C) = (A \cap B) \cup (A \cap C)$.

EXERCISE 2.3

1. Prove that $A \subseteq B \Leftrightarrow A \cup B = B$.
2. Prove that $A \subseteq B \Leftrightarrow A \cap B = A$.
3. Prove parts (a), (b), and (c) of Theorem 4.
4. Prove the second half of Theorem 4(d).
5. Prove the second half of Theorem 4(e).
6. Prove that
   (a) $A \subseteq C$ and $B \subseteq C$ implies $(A \cup B) \subseteq C$
   (b) $A \subseteq B$ and $A \subseteq C$ implies $A \subseteq (B \cap C)$.
   [Hint: Use Theorem 5 in Chapter 1, if you wish.]
7. Prove that $(A \cap B) \cup C = A \cap (B \cup C) \Leftrightarrow C \subseteq A$.

8. Prove that if $A \subseteq B$ then $\mathscr{P}(A) \subseteq \mathscr{P}(B)$.
9. Prove that $A \cup B = A \cap B \Leftrightarrow A = B$.
10. Prove that if $A \subseteq B$, then $A \cup C \subseteq B \cup C$ and $A \cap C \subseteq B \cap C$ for any set C.
11. Prove that if $A \subseteq C$ and $B \subseteq D$, then $A \cup B \subseteq C \cup D$.
12. Prove or disprove that
    (a) If $A \cup B = A \cup C$ then $B = C$.
    (b) If $A \cap B = A \cap C$ then $B = C$.
13. Prove that if $A, B_1, B_2, \ldots, B_n$ are sets, then
    (a) $A \cap (B_1 \cup B_2 \cup \cdots \cup B_n) = (A \cap B_1) \cup (A \cap B_2) \cup \cdots \cup (A \cup B_n)$
14. Prove that
    (a) If $A_i \subseteq B$ for $i = 1, 2, \ldots, n$, then $A_1 \cup A_2 \cup \cdots \cup A_n \subseteq B$.
    (b) If $A \subseteq B_i$ for $i = 1, 2, \ldots, n$, then $A \subseteq B_1 \cap B_2 \cap \cdots \cap B_n$.

4. COMPLEMENTS

There is, in set theory, an operation known as complementation, which is similar to the operation of subtraction in arithmetic.

Definition 5. If A and B are sets, the *relative complement* of B in A is the set $A - B$ defined by

$$A - B = \{x \in A \mid x \notin B\}$$

In this definition it is *not* assumed that $B \subseteq A$.

EXAMPLE 4. Let

$$A = \{a, b, c, d\} \qquad \text{and} \qquad B = \{c, d, e, f\}$$

Find $A - B$ and $A - (A \cap B)$.

Solution.

$$A - B = \{a, b, c, d\} - \{c, d, e, f\} = \{a, b\}$$

and

$$A - (A \cap B) = \{a, b, c, d\} - \{c, d\} = \{a, b\}$$

Although the universal set in the absolute sense, the set of all sets, does not exist (see the Russell Paradox in Section 7), it does no harm to assume temporarily that *all the sets to be mentioned in the rest of this book are subsets of a fixed set U* which may be regarded (temporarily) as a universal set in a restricted sense. In order to state the basic rules concerning complementation as simply as possible, we assume, unless otherwise stated, that all complements are formed relative to this set U. We shall then write A' for $U - A$.

EXAMPLE 5. Show that $A - B = A \cap B'$.

Solution.

$x \in A \cap B'$	$\equiv (x \in A) \wedge (x \in U - B)$	Def. $\cap$, Def. of $'$
	$\equiv (x \in A) \wedge [(x \in U) \wedge (x \notin B)]$	Def. 5
	$\equiv [(x \in A) \wedge (x \in U)] \wedge (x \notin B)$	Assoc.
	$\equiv (x \in A \cap U) \wedge (x \notin B)$	Def. $\cap$
	$\equiv (x \in A) \wedge (x \notin B)$	$A \cap U = A$
	$\equiv x \in (A - B)$	Def. 5

Therefore, by Definition 1, $A \cap B' = A - B$.

Theorem 5. Let A and B be sets. Then

(a) $(A')' = A$.

(b) $\varnothing' = U$ and $U' = \varnothing$.

(c) $A \cap A' = \varnothing$ and $A \cup A' = U$.

(d) $A \subseteq B$ if and only if $B' \subseteq A'$.

Proof. The proofs for parts (a), (b), (c) use only the definitions and are left to the reader as exercises. We give a proof for part (d):

$A \subseteq B$	$\equiv [(x \in A) \rightarrow (x \in B)]$	Def. $\subseteq$
	$\equiv [(x \notin B) \rightarrow (x \notin A)]$ [4]	Contrap.
	$\equiv [(x \in B') \rightarrow (x \in A')]$	Def. of $'$
	$\equiv (B' \subseteq A')$	Def. $\subseteq$

We have thus proved that $(A \subseteq B) \equiv (B' \subseteq A')$.

In the above proof, again symbols and laws of logic (from Chapter 1) are used, which enable us to display each step of the proof neatly and precisely with supporting reasons on the right-hand side. The reader is encouraged to make full use of Chapter 1 for proofs whenever possible.

The most useful property of complements is the following De Morgan's Theorem. The reader should compare this theorem with De Morgan's Law in Chapter 1.

---

[4] Recall that the negation of $x \in B$, $\sim(x \in B)$, is denoted by $x \notin B$.

Theorem 6. (*De Morgan's Theorem*). For any sets A and B,

(a) $(A \cup B)' = A' \cap B'$.

(b) $(A \cap B)' = A' \cup B'$.

Proof (a):

$x \in (A \cup B)' \equiv \sim[x \in A \cup B)]$	Def. of $'$
$\equiv \sim[(x \in A) \vee (x \in B)]$	Def. $\cup$
$\equiv \sim(x \in A) \wedge \sim(x \in B)$	De M. of logic
$\equiv (x \in A') \wedge (x \in B')$	Def. of $'$
$\equiv x \in (A' \cap B')$	Def. $\cap$

Therefore, by Definition 1, $(A \cup B)' = A' \cap B'$.

The proof for (b) is left to the reader.

EXAMPLE 6. Let A, B, and C be any sets. Decide whether the set $A \cap (B - C)$ is the same as the set $(A \cap B) - (A \cap C)$.

Solution.

$(A \cap B) - (A \cap C) = (A \cap B) \cap (A \cap C)'$	Example 5
$= (A \cap B) \cap (A' \cup C')$	De M. Th. (Th. 6)
$= (A \cap B \cap A') \cup (A \cap B \cap C')$	Dist.
$= (A \cap A' \cap B) \cup (A \cap B \cap C')$	Com.
$= \varnothing \cup [A \cap (B \cap C')]$	Th. 5(c): $A \cap A' = \varnothing$
$= A \cap (B - C)$	Th. 4(a) Example 5

Hence, we have proved that $A \cap (B - C) = (A \cap B) - (A \cap C)$.

EXERCISE 2.4

1. Let A and B be sets. Prove that $A - B = A - (B \cap A)$.
2. Prove parts (a), (b), and (c) of Theorem 5.
3. Let A and B be sets. Prove that $B \subseteq A'$ if and only if $A \cap B = \varnothing$.

4. Let A and B be sets. Prove that $A \cap B = \varnothing$ if and only if $A \cap B' = A$.
5. Prove that for any sets A and B, $A' - B' = B - A$.
6. Let A and B be sets. Prove that $(A - B) \cup B = A$ if and only if $B \subseteq A$.
7. Prove Theorem 6(b).
8. Let A, B, and C be any sets. Prove that
   (a) $(A - C) \cup (B - C) = (A \cup B) - C$
   (b) $(A - C) \cap (B - C) = (A \cap B) - C$.
9. Let $A_1, A_2, \ldots, A_n$ and C be sets. Prove that
   $(A_1 - C) \cup (A_2 - C) \cup \cdots \cup (A_n - C) = (A_1 \cup A_2 \cup \cdots \cup A_n) - C$.
10. Let $B_1, B_2, \ldots, B_n$ and C be sets. Prove that
   $(B_1 - C) \cap (B_2 - C) \cap \cdots \cap (B_n - C) = (B_1 \cap B_2 \cap \cdots \cap B_n) - C$.
11. Let A and B be sets. Prove that A and $B - A$ are disjoint and that $A \cup B = A \cup (B - A)$. (This show how to represent the union $A \cup B$ as a disjoint union.)
12. Prove that $A \cup B = (A' \cap B')'$ and $A \cap B = (A' \cup B')'$.
13. Under what conditions on sets A and B, does $A - B = B - A$?
14. Let A, B, and C be sets. Prove that
$$(A - B) - C = (A - C) - (B - C).$$
15. Let A, B, and C be sets. Prove that
$$(A - B) - C = A - (B \cup C).$$
16. Let A and B be sets. Prove that $A \cap B' = \varnothing = A' \cap B$ if and only if $A = B$.
17. Let A and B be subsets of a set X. Prove or disprove that $(X - A) \cap (X - B) = X - (A \cap B)$.
18. Let A, B, and C be any sets. Prove that
   (a) $(A \cap B \cap C)' = A' \cup B' \cup C'$
   (b) $(A \cup B \cup C)' = A' \cap B' \cap C'$.
   Generalize these results to statements involving sets
$$A_1, A_2, A_3, \ldots, A_n.$$
19. For any sets A and B, prove or disprove that
   (a) $\mathscr{P}(A) \cap \mathscr{P}(B) = \mathscr{P}(A \cap B)$
   (b) $\mathscr{P}(A) \cup \mathscr{P}(B) = \mathscr{P}(A \cup B)$.
20. Prove that if $A \subseteq C, B \subseteq C, A \cup B = C$, and $A \cap B = \varnothing$, then $A = C - B$.
21. Let A and B be any sets. Prove that
$$(A - B) \cup (B - A) = (A \cup B) - (A \cap B).$$
22. Let X be the power set of a given nonempty set U. In X we define a binary operation $\oplus$, called the *symmetric difference*, by
$$A \oplus B = (A - B) \cup (B - A)$$
   for all A and B in X. Prove that for all A, B, and C in X.

(a) $A \oplus B \in X$
(b) $A \oplus B = B \oplus A$
(c) $A \oplus (B \oplus C) = (A \oplus B) \oplus C$
(d) $A \oplus \varnothing = A$
(e) $A \oplus A = \varnothing$.

23. Under the assumptions of Problem 22, above, prove that
(a) $A \oplus A' = U$
(b) $A \oplus C = A$ if and only if $C = \varnothing$
(c) $A \oplus C = \varnothing$ if and only if $C = A$
(d) If $A \oplus C = B \oplus C$ then $A = B$.
24. Under the assumptions of Problem 22, above, prove that
$$A \cap (B \oplus C) = (A \cap B) \oplus (A \cap C)$$

Remark: A system consists of a set together with a binary operation, such as $(X, \oplus)$, satisfying condition (a) to (e) of Problem 22 is said to form an (algebraic) *group*.

5. VENN DIAGRAMS

To help in visualizing set operations and their results we introduce diagrams, called Venn diagrams, which illustrate them. We shall represent the imaginary relative universal set U by a rectangle and let subsets (of U) be circles drawn inside the rectangle. For example, in Figure 1 we represent

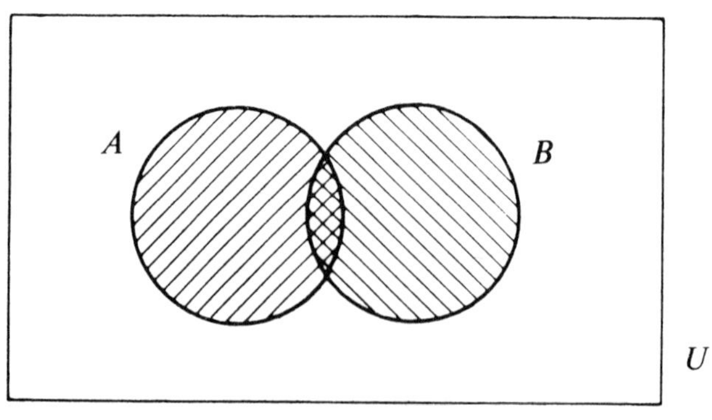

Figure 1.

two sets A and B as two shaded circles; the doubly crosshatched part is the intersection $A \cap B$ and the total shaded area is the union $A \cup B$.

Figure 2 shows two sets A and B that are disjoint. The shaded area in Figure 3 represents the complement A' of the set A. The set $A - B$, the relative complement of B in A, is represented by the shaded part in Figure 4.

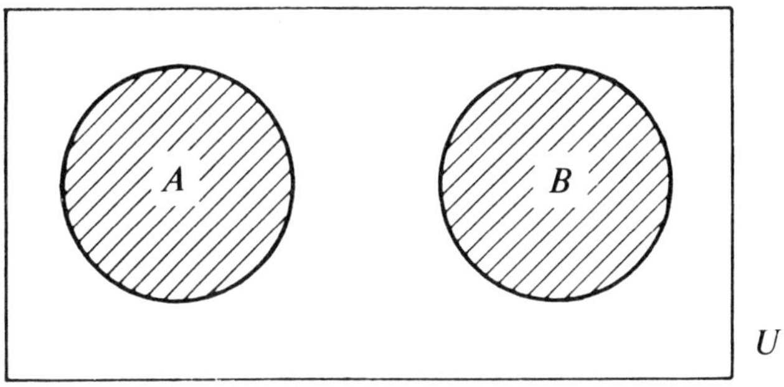

Figure 2.

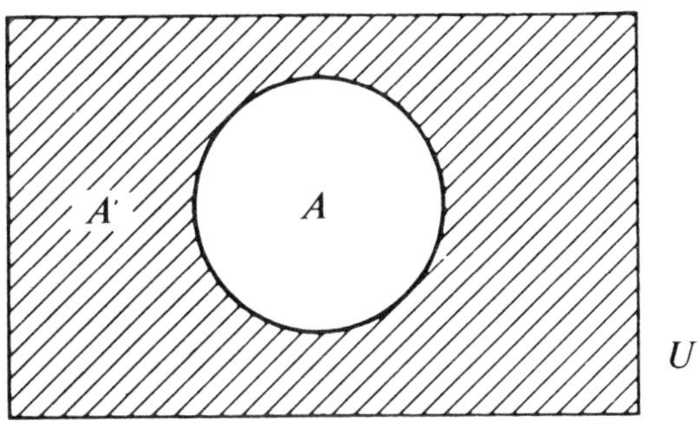

Figure 3.

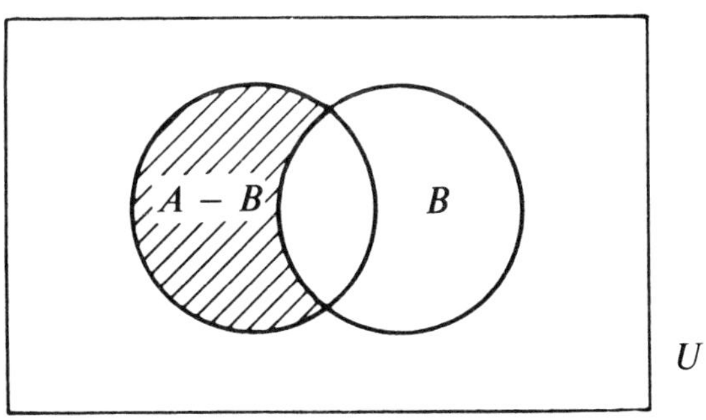

Figure 4.

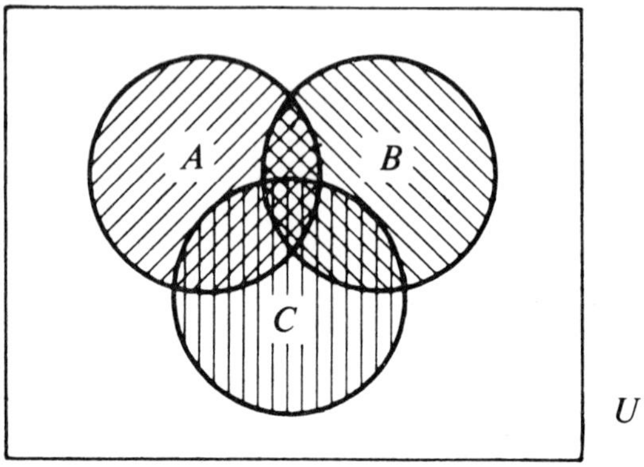

Figure 5.

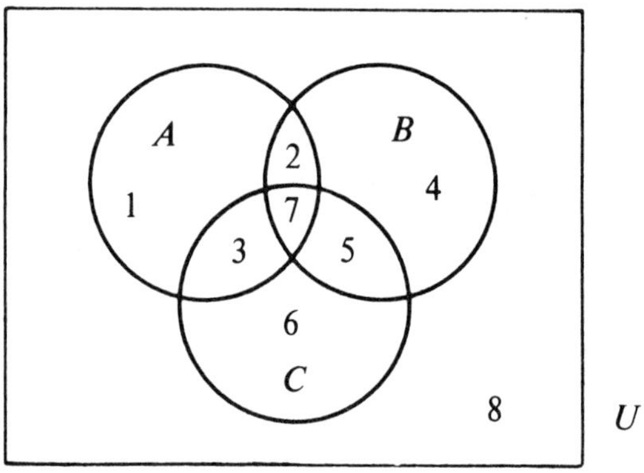

Figure 6.

A typical Venn diagram of three sets A, B, and C may be drawn as Figure 5. These three sets divide the universal set U into 8 parts as indicated in Figure 6.

Using the above diagram, we can give a simple heuristic argument for the validity of, for example, the distributive law, $A \cap (B \cup C) = (A \cap B) \cup (A \cap C)$, as follows: From Figure 6, $A \cap (B \cup C)$ consists of areas 2, 3, and 7. On the other hand, $(A \cap B) \cup (A \cap C)$ is represented by the union of areas 2 and 7, and areas 3 and 7. Therefore, the equality $A \cap (B \cup C) = (A \cap B) \cup (A \cap C)$ seems plausible. However, in mathematics a heuristic argument cannot be accepted as a proof.

EXERCISE 2.5

1. Draw a Venn diagram for $A \subset B$.
2. Draw Venn diagrams for $A \cap B'$, $A' \cap B$, and $A' \cap B'$.
3. Draw Venn diagrams for $A \cup B'$, $A' \cup B$, and $A' \cup B'$.
4. Draw a Venn diagram for $A \oplus B$.

In Problems 5 through 13, draw Venn diagrams and give Heuristic arguments that each of these statements is plausable.

5. $A \cap (B \cap C) = (A \cap B) \cap C$
6. $A \cup (B \cup C) = (A \cup B) \cup C$
7. $A \cup (B \cap C) = (A \cup B) \cap (A \cup C)$
8. $(A \cup B)' = A' \cap B'$
9. $(A \cap B)' = A' \cup B'$
10. $A \cap (B - A) = \varnothing$ and $A \cup (B - A) = A \cup B$
11. $(A \cup B) - (A \cap B) = (A - B) \cup (B - A)$
12. $A \oplus (B \oplus C) = (A \oplus B) \oplus C$ [Hint: for 12-14, use figure 6.]
13. $A \cap (B \oplus C) = (A \cap B) \oplus (A \cap C)$

In Problems 14 through 20, draw Venn diagrams to disprove the following:

14. $A \oplus (B \cap C) = (A \oplus B) \cap (A \oplus C)$
15. If $x \in A$ and $A \not\subset B$, then $x \notin B$. [Prob. 8(e), Ex. 2.1]
16. If $A \not\subset B$ and $B \not\subset C$, then $A \not\subset C$. [Prob. 8(d), Ex. 2.1]
17. If $A \not\subset B$ and $B \not\subset C$, then $A \not\subset C$. [Prob. 8(c) Ex. 2.1]
18. (a) If $A \cup B = A \cup C$ then $B = C$. [prob. 12, Ex. 2.3]
    (b) If $A \cap B = A \cap C$ then $B = C$.
19. If A and B are subsets of X, then
    $(X - A) \cap (X - B) = X - (A \cap B)$.[Prob. 17, Ex. 2.4]
20. If A and B are subsets of X, then $(X - A) \cup (X - B) = X - (A \cup B)$.

6. INDEXED FAMILIES OF SETS

Recall that a set is a collection of elements that are all distinct. Roughly speaking a *family* is to be considered as a collection of not necessarily distinct objects called *members*. For example, $\{a, a, a\}$ is a family with three members a, a, and a. But the same family $\{a, a, a\}$ considered as a set is just the singleton set $\{a\}$ with only one element, a.

Let Γ be a set and assume that with each element γ of Γ there is assiciated a set A_γ. The family of all such sets A_γ is called an *indexed family of sets* indexed by the set Γ and is denoted by

$$\{A_\gamma \mid \gamma \in \Gamma\}$$

For example, the family of sets: $\{1, 2\}, \{2, 4\}, \{3, 6\}, \ldots, \{n, 2n\} \ldots$ may be considered as an indexed family of sets indexed by the set $\mathbf{N}$ of natural numbers, where $A_n = \{n, 2n\}$ for each $n \in \mathbf{N}$. This family of sets may be denoted by $\{\{n, 2n\} \mid n \in \mathbf{N}\}$.

An arbitrary family of sets might not appear to be indexed, but in most cases one can easily find a set Γ which can be used to index the given family of sets.

EXAMPLE 7. Index the family $\mathcal{F}$ of sets $\varnothing$, **N**, **Z**, **Q**, **R**, and **R**.

Solution. Since this family contains exactly six members (although two of them are the same), we choose $\Gamma = \{1, 2, 3, 4, 5, 6\}$ and let $A_1 = \varnothing$, $A_2 = \mathbf{N}$, $A_3 = \mathbf{Z}$, $A_4 = \mathbf{Q}$, $A_5 = \mathbf{R}$, and $A_6 = \mathbf{R}$. The family of sets is then indexed.

Virtually all symbols and notations used for sets will apply to families as well. For instance, $\varnothing \in \mathcal{F}$ and $\mathbf{R}_+ \notin \mathcal{F}$ will indicate respectively that $\varnothing$ is a member of the family $\mathcal{F}$ and $\mathbf{R}_+$ is not a member of $\mathcal{F}$. We may also write $\mathcal{F} = \{\varnothing, \mathbf{N}, \mathbf{Z}, \mathbf{Q}, \mathbf{R}, \mathbf{R}\}$.

Let us now extend the concepts of union $\cup$ and intersection $\cap$ defined in Definitions 3 and 4 to an arbitrary family of sets.

Definition 6. Let $\mathcal{F}$ be an arbitrary family of sets. The *union* of the sets in $\mathcal{F}$, denoted by $\bigcup_{A \in \mathcal{F}} A$ or $\bigcup \mathcal{F}$, is the set of all elements that are in A for some $A \in \mathcal{F}$. That is,

$$\bigcup_{A \in \mathcal{F}} A = \{x \in U \mid x \in A \text{ for some } A \in \mathcal{F}\}$$

If the family $\mathcal{F}$ is indexed by the set Γ, the following alternate notation may be used:

$$\bigcup_{\gamma \in \Gamma} A_\gamma = \{x \in U \mid x \in A_\gamma \text{ for some } \gamma \in \Gamma\}$$

If the index set Γ is finite, $\Gamma = \{1, 2, 3, \ldots, n\}$ for some natural number n, more intuitive notations such as

$$\bigcup_{i=1}^{n} A_i \qquad \text{or} \qquad A_1 \cup A_2 \cup \cdots \cup A_n.$$

are often used for $\bigcup_{\gamma \in \Gamma} A_\gamma$.

EXAMPLE 8. Find the union of the family of sets

$$\{1\}, \{2,3\}, \{3, 4, 5\}, \ldots, \{n, n + 1, \ldots, 2n - 1\}.$$

Solution. This family of sets may be considered as indexed by $\Gamma = \{1, 2, 3, \ldots, n\}$, where $A_i = \{i, i + 1, \ldots, 2i - 1\}$ for each $i \in \Gamma$. The problem reduces to that of finding $\bigcup_{i=1}^{n}\{i, i + 1, \ldots, 2i - 1\}$. Observe that each integer between 1 and $2n - 1$ belongs to some A_i in the family, and no other element belongs to any of these A_i. hence,

$$\bigcup_{i=1}^{n} \{i, i + 1, \ldots, 2i - 1\} = \{1, 2, 3, \ldots, 2n - 1\}$$

Definition 7. Let $\mathscr{F}$ be an arbitrary family of sets. The *intersection* of sets in $\mathscr{F}$, denoted by $\bigcap_{A \in \mathscr{F}} A$ or $\bigcap \mathscr{F}$, is the set of all elements that are in A for all $A \in \mathscr{F}$. That is,

$$\bigcap_{A \in \mathscr{F}} A = \{x \in U \mid x \in A \text{ for all } A \in \mathscr{F}\}$$

Here, the statement "$x \in A$ for all $A \in \mathscr{F}$" may be expressed alternately as "$A \in \mathscr{F} \rightarrow x \in A$." The latter expression has an advantage in proving theorems, as we shall see in the coming Theorem 7.

If the family $\mathscr{F}$ is indexed by the set Γ, the following alternate notation may be used:

$$\bigcap_{\gamma \in \Gamma} A_\gamma = \{x \in U \mid x \in A_\gamma \text{ for all } \gamma \in \Gamma\}$$

If the index set Γ is finite, $\Gamma = \{1, 2, \ldots, n\}$ for some positive integer n, then as in the case of union we usually write

$$\bigcap_{i=1}^{n} A_i \qquad \text{or} \qquad A_1 \cap A_2 \cap \cdots \cap A_n$$

instead of $\bigcap_{\gamma \in \Gamma} A_\gamma$.

Let a and b be any two real numbers. By an *open interval* (a, b) we mean the subset $\{x \in \mathbf{R} \mid a < x < b\}$ of $\mathbf{R}$. It follows that if $a \geq b$ then the interval $(a, b) = \varnothing$.

EXAMPLE 9. Find the intersection of the family of open intervals

$$(0, 1), (0, \tfrac{1}{2}), (0, \tfrac{1}{3}), \ldots$$

Solution. We are to find the set $\bigcap_{n \in \mathbf{N}} (0, 1/n)$. Intuitively speaking, the given family is a sequence of "decreasing" intervals $(0, 1/n)$, where the interval $(0, 1/n)$ "approaches" the empty set $\varnothing$ as n becomes large. Therefore, we may conjecture that the intersection $\bigcap_{n \in \mathbf{N}} (0, 1/n)$ should be the empty set. We now prove that this conjecture is true. Suppose on the contrary that there exists some real number $a \in \bigcap_{n \in \mathbf{N}} (0, 1/n)$. Then we would have $0 < a < 1/n$ for all $n \in \mathbf{N}$. This contradicts the fact that for a fixed $a > 0$

there always exists a sufficiently large n in $\mathbf{N}$ such that $1/n < a$. The contradiction shows that $\bigcap_{n\in\mathbf{N}}(0,1/n) = \varnothing$.

Theorem 7. Let $\{A_\gamma \mid \gamma\in\Gamma\}$ be the empty family of sets; that is, $\Gamma = \varnothing$. Then
(a) $\bigcup_{\gamma\in\varnothing}A_\gamma = \varnothing$.
(b) $\bigcap_{\gamma\in\varnothing}A_\gamma = U$.

Proof. (a) To show $\bigcup_{\gamma\in\varnothing}A_\gamma = \varnothing$, we show equivalently that $x\notin\bigcup_{\gamma\in\varnothing}A_\gamma$ for all x (in U):

$$\begin{aligned} x\notin \bigcup_{\gamma\in\varnothing} A_\gamma &\equiv \sim\left(x\in \bigcup_{\gamma\in\varnothing} A_\gamma\right) && \text{Notation}\\ &\equiv \sim(x\in A_\gamma \text{ for some } \gamma\subset\varnothing) && \text{Def. 6}\\ &\equiv (x\notin A_\gamma) \text{ for all } \gamma\in\varnothing] && \text{Q.N. (Ch. 1)}\\ &\equiv (\gamma\in\varnothing\rightarrow x\notin A_\gamma) \end{aligned}$$

The last statement is, by Theorem 7(c) of Chapter 1, true for all $x\in U$, because $\gamma\in\varnothing$ is a contradiction. This completes the proof of part (a).

(b) We shall prove that $x\in\bigcap_{\gamma\in\varnothing}A_\gamma$ for all x in U. Observe that

$$\begin{aligned} x\in \bigcap_{\gamma\in\varnothing} A_\gamma &\equiv (x\in A_\gamma\,\forall\gamma\in\varnothing) && \text{Def. 7}\\ &\equiv (\gamma\in\varnothing\rightarrow x\in A_\gamma) \end{aligned}$$

The last assertion is, as we have explained in the proof of part (a), a true statement for all $x\in U$. The proof is complete.

Many theorems concerning operations of finitely many sets can be generalized to theorems concerning operations of an arbitrary family of sets. For example, the following generalizes De Morgan's Theorem. The student should compare this theorem with Theorem 6.

Theorem 8. (*The Generalized De Morgan Theorem*). Let $\{A_\gamma \mid \gamma\in\Gamma\}$ be an arbitrary family of sets. Then

(a) $(\bigcup_{\gamma\in\Gamma}A_\gamma)' = \bigcap_{\gamma\in\Gamma}A'_\gamma$.
(b) $(\bigcap_{\gamma\in\Gamma}A_\gamma)' = \bigcup_{\gamma\in\Gamma}A'_\gamma$.

Proof. We shall only prove part (a), and leave part (b) to the student.

$$\begin{aligned} x\in\left(\bigcup_{\gamma\in\Gamma}A_\gamma\right)' &\equiv \sim\left(x\in\bigcup_{\gamma\in\Gamma}A_\gamma\right) && \text{Def. of}\,'\\ &\equiv \sim(\exists\gamma\in\Gamma)(x\in A_\gamma) && \text{Def. 6}\\ &\equiv (\forall\gamma\in\Gamma)(x\notin A_\gamma) && \text{Q.N. (Ch. 1)} \end{aligned}$$

$\equiv (\forall\gamma\in\Gamma)\,(x\in A'_\gamma)$ Def. of $'$

$\equiv x\in \bigcap_{\gamma\in\Gamma} A'_\gamma$ Def. 7

Therefore, by Definition 1, $(\bigcup_{\gamma\in\Gamma} A)' = \bigcap_{\gamma\in\Gamma} A'_\gamma$.

The following theorem is a generalization of Theorem 4(e).

Theorem 9. (*Generalized Distributive Laws*). Let A be a set and let $\mathscr{F} = \{B_\gamma \mid \gamma\in\Gamma\}$ be an arbitrary family of sets. Then

(a) $A \cap (\bigcup_{\gamma\in\Gamma} B_\gamma) = \bigcup_{\gamma\in\Gamma} (A \cap B_\gamma)$.
(b) $A \cup (\bigcap_{\gamma\in\Gamma} B_\gamma) = \bigcap_{\gamma\in\Gamma} (A \cup B_\gamma)$.

Proof. (a) An element x is in the set $A\cap(\bigcup_{\gamma\in\Gamma} B_\gamma)$ if and only if $x\in A$ and $x\in\bigcup_{\gamma\in\Gamma} B_\gamma$, which according to Definition 6, is equivalent to

$$x\in A \quad \text{and} \quad x\in B_\gamma \quad \text{for some} \quad \gamma\in\Gamma$$

The last assertion may be expressed, by Definition 4, as

$$x\in A\cap B_\gamma \quad \text{for some} \quad \gamma\in\Gamma$$

which by Definition 6 is precisely $x\in\bigcup_{\gamma\in\Gamma}(A\cap B_\gamma)$. Thus, by Definition 1, $A\cap(\bigcup_{\gamma\in\Gamma} B_\gamma) = \bigcup_{\gamma\in\Gamma} (A\cap B_\gamma)$.

The proof for part (b) is an exercise.

EXERCISE 2.6

1. Let $\Gamma = \{1, 2, 3, 4\}$ and $A_1 = \{a, b, c, d\}, A_2 = \{b, c, d\}, A_3 = \{a, b, c\}, A_4 = \{a, b\}$. Find the following.
   (a) $\bigcup_{i=1}^{4} A_i$
   (b) $\bigcap_{i=1}^{4} A_i$
2. For any real numbers a and b, by the *closed interval* $[a, b]$ we mean the set $\{x\in\mathbf{R} \mid a \le x \le b\}$. If $a > b, [a, b] = \varnothing$. Find the following sets.
   (a) $\bigcap_{n\in\mathbf{N}} [0, 1/n]$
   (b) $\bigcup_{n\in\mathbf{N}} [0, 1/n]$
   (c) $\bigcap_{n=1}^{99} [0, 1/n]$
3. Let A_k be the open interval $(-1/k, 1/k)$ for each $k\in\mathbf{N}$. Find the following:
   (a) $\bigcup_{k\in\mathbf{N}} A_k$
   (b) $\bigcap_{k\in\mathbf{N}} A_k$.
4. For any real numers a and b, by the *half closed half open* intervals $(a, b]$ and $[a, b)$ we mean respectively the set $\{x\in\mathbf{R} \mid a < x \le b\}$ and

$\{x \in \mathbf{R} \mid a \leq x < b\}$. Find the following sets:
(a) $\bigcup_{k \in \mathbf{N}} (k, k + 1]$
(b) $\bigcup_{k \in \mathbf{N}} (-k, -k + 1]$
(c) $\bigcup_{k \in \mathbf{N}} [k, k + 1)$
(d) $\bigcup_{k \in \mathbf{N}} [-k, k)$.

5. Prove Theorem 8(b): $(\bigcap_{\gamma \in \Gamma} A_\gamma)' = \bigcup_{\gamma \in \Gamma} A'_\gamma$.
6. Prove Theorem 9(b): $A \cup (\bigcap_{\gamma \in \Gamma} B_\gamma) = \bigcap_{\gamma \in \Gamma} (A \cup B)$.
7. Expand
   (a) $(A_1 \cup A_2) \cap (B_1 \cup B_2 \cup B_3)$
   into a union of intersections, and
   (b) $(A_1 \cap A_2) \cup (B_1 \cap B_2 \cap B_3)$
   into an intersection of unions. [Hint: Use Theorem 9 several times.]
8. Expand
   (a) $(\bigcup_{i=1}^{m} A_i) \cap (\bigcup_{j=1}^{n} B_j)$
   into a union of intersections, and
   (b) $(\bigcap_{i=1}^{m} A_i) \cup (\bigcap_{j=1}^{n} B_j)$
   into an intersection of unions. [See Problems 5.]
9. Let $\{A_\gamma \mid \gamma \in \Gamma\}$ and $\{B_\delta \mid \delta \in \Delta\}$ be any two families of sets. expand
   (a) $(\bigcup_{\gamma \in \Gamma} A_\gamma) \cap (\bigcup_{\delta \in \Delta} B_\delta)$
   into a union of intersections, and
   (b) $(\bigcap_{\gamma \in \Gamma} A_\gamma) \cup (\bigcap_{\delta \in \Delta} B_\delta)$
   into an intersection of unions. [See Problems 5 and 6.]

7. THE RUSSELL PARADOX

Now many of us may think we understand what is meant by a set—at least intuitively. Most of us taking a Set Theory course for the first time would not notice what is wrong in considering "the set of all sets" or the so-called "universal set in the absolute sense. In fact, for a peroid of time (at least from 1895 when Georg Cantor first created a theory of sets, until 1902 when the Russell Paradox appeared), the existence of such a universal set was taken for granted. It was the famous English philosopher Bertrand Russell (1872-1970)[5] who shook the mathematics community in 1902 by

[5] Bertrand Russell was born on May 18, 1872, at Trelleck, Wales. Before he was four, both of his parents died. He had been a shy, silent boy until he entered Trinity College, Cambridge University, in 1890. After three years of mathematics he concluded that what he was being taught was full of errors. He sold his mathematics books and changed to philosophy. In his *Principia Mathematica* (1910-1913), a three-volume monumental work co-authored with Alfred North Whitehead (1861-1947), he attempted to recast set theory so as to avoid paradoxes. In 1918 he wrote, "I want to stand at the rim of the world and peer into the darkness beyond, and see a little more than others have seen. . . . I want to bring back into the world of men some little bit of wisdom." He certainly did, more than just "some little bit." In the same year, he was put in prison for unfavorable comment about the American Army. In 1950 he received the Order of Merit from the King of England and the Nobel prize for literature. In his late years he led a number of demonstrations against nuclear warfare.

declaring that the admission of a set of all sets would lead to a contradiction. This is the famous Russell Paradox. We present this paradox as two seemingly contradictory lemmas from which a theorem follows.

Lemma 1. Suppose that there is a set $\mathcal{U}$ of all sets. Let $R = \{S \in \mathcal{U} \mid S \notin S\}$.[6] Then $R \notin R$.

Proof. Suppose on the contrary the $R \in R$. Then by the specification of the set R, we must have $R \notin R$, which contradicts the assumption that $R \in R$. The contradiction proves that $R \notin R$.

Lemma 2. Suppose that there is a set $\mathcal{U}$ of all sets. Let R be the set $\{S \in \mathcal{U} \mid S \notin S\}$. Then $R \in R$.

Proof. Suppose the contrary, that $R \notin R$. Then since $R \in \mathcal{U}$, we have $R \in R$ by the definition of R. This is a contradiction. Thus, $R \in R$.

Theorem 10. There does not exist a set of all sets.

Proof. In view of Lemmas 1 and 2, the set of all sets cannot exist. For, if it were otherwise, it would lead to the contradiction "$R \notin R$ and $R \in R$."

Paul R. Halmos puts it this way: "*Nothing contains everything.*"[7]

8. HISTORICAL REMARK

The modern theory of sets is generally considered to heve been created in 1895 by the famous mathematican Georg Cantor[8] (1845-1918), who noticed the need for such a theory while studying trigonometric series. Cantor wrote: "By a 'set' we shall understand any collection into a whole of definite distinct objects of our intuition or thought." This definition does not prohibit anyone from considering the "set" of all sets, as Bertrand Russell did. The real difficulty in Cantor's definition of a set is the word "collection." What is a collection? Of course we can look it up in a dictionary and find something like these definitions:

---

[6] According to the rule of specification, R is a set which is often called "the Russell set."

[7] Paul R. Halmos, *Naive Set Theory*, D. Van Nostrand Company, Inc., New York, 1960, p.6.

[8] Georg Cantor was born in St. Petersburg, Russia in 1845, moved to Germany in 1856, studied mathematics at the University of Berlin (1863-1869), and taught at the University of Halle (1869-1905). One of Cantor's interests was trigonometric series, which led him to look into the foundation on analysis. As a result, he created the revolutionary work on set theory and an arithmetic of transfinite numbers.

"collection: a group of collected objects."

"group: an aggregate or collection."

"aggregate: a collection."

These will hardly be of any help. When a mathemtician gives a definition it is not intended to be a mere synonym such as "collection" for "set," of a circular definition as we would find in a dictionary. Cantor apparently was not aware that the term "set" was really undefinable.

To avoid any difficulty such as the Russell Paradox in set theory, we must accept the terms "set" and "element" as undefined terms, or primitives, and guide these premitives by a number of axioms, including the Axiom of Specification and the Axiom of Power Sets that have been introduced in Section 2. Other axioms such as "$A = B$ if and only if A and B contain the same elements" (Axiom of Extension), "$\varnothing$ is a set" (Axiom of the Empty Set), "If A and B are sets then so is $\{A,B\}$" (Axiom of Pairing), and "If $\mathscr{F}$ is a set of sets then $\bigcup \mathscr{F}$ is a set" (Axiom of Unions) are often given in axiomatic treatments of set theory.

The Russell Paradox was not the only one to arise in set theory. Shortly after the Russell Paradox appeared, many paradoxes were constructed by several mathematicians and logicians. As a consequence of all these paradoxes, many mathematicians and logicians have contributed to several brands of "axiomatic set theory," each designed to avoid these paradoxes and, at the same time, to preserve the main body of Cantor's set theory. However, at the time of this writing, no one yet come up with a completely satisfactory axiomatic system for the set theory.

Despite the aforementioned difficulties, Cantor's set theory has today penetrated into every branch of modern mathematics, and it has proved to be of particular importance in the foundations of modern analysis and in topology. In fact, even the very simplest full-fledged axiomatic systems of set theory are entirely adequate for doing virtually all of classical mathematics (e.g. the theory of real and complex numbers, algebra, topology, etc.).

Chapter Three

Relations and Functions

The chapter begins with a discussion of ordered pairs and the Cartesian product of two sets. A relation is then defined as a set of ordered pairs. The intimate connection between a partition and an equivalence relation on a set is closely examined. The concept of a function is introduced as a special kind of relation. As a preparation for those readers who wish to pursue more modern mathematics, important properties of functions are studied. An abundance of examples is provided.

1. CARTESIAN PRODUCT OF TWO SETS

Given any two objects a and b, we may form a new object (a, b), called the *ordered pair a, b.*[1] The adjective "ordered" here emphasizes that the order in which the objects a and b appear in the brackets is essential. Thus, (a, b) and (b, a) are two distinct ordered pairs. it should be noted that an ordered pair (a, b) is not the same as the set $\{a, b\}$. There is a satisfactory, but somewhat cumbersome, way of defining the ordered pair (a, b) as the set $\{\{a\}, \{a, b\}\}$, from which the property "$(a, b) = (c, d) \Leftrightarrow a = c$ and $b = d$" follows. (See Problem 21, Exercise 3.1.)

Two ordered pairs (a, b) and (c, d) are said to be equal $(=)$ if and only if $a = c$ and $b = d$. For example, $(x, y) = (7, 8)$ if and only if $x = 7$ and $y = 8$.

In analytic geometry, the Cartesian plane may be considered as the set of all ordered pairs of real numbers. We state this concept formally as follows:

[1] Unfortunately the notation (a, b) for an ordered pair is the same as the notation for an open interval when a and b are real numbers. However, the careful reader should always be able to make the distinction from the context.

Definition 1. Let A and B be two sets. The set of all ordered pairs (x, y), with $x \in A$ and $y \in B$, is called the *Cartesian product* of A and B, and is denoted by $A \times B$. In symbols

$$A \times B = \{(x, y) \mid x \in A \wedge y \in B\}$$

For the ordered pair (a, b), a is called the *first coordinate* and b is the *second coordinate.*

EXAMPLE 1. Let $A = \{a, b, c\}$ and $B = \{1, 2\}$. Find the Cartesian products $A \times B$ and $B \times A$.

Solutions. By Definition 1 above, we have

$$A \times B = \{(a, 1), (a, 2), (b, 1), (b, 2), (c, 1), (c, 2)\}$$

and

$$B \times A = \{(1, a), (1, b), (1, c), (2, a), (2, b), (2, c)\}$$

We notice that $A \times B \neq B \times A$. We may picture the Cartesian product $A \times B$ as the set of dots in the following figure.

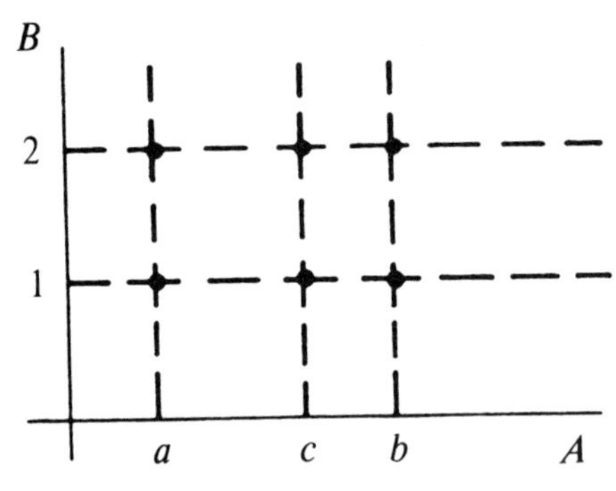

Figure 7.

EXAMPLE 2. Let A be any set. Find $A \times \varnothing$ and $\varnothing \times A$.

Solution. Since $A \times \varnothing$ is the set of all ordered pairs (a, b) such that $a \in A$ and $b \in \varnothing$, and since the empty set $\varnothing$ contains no elements, there is no b in $\varnothing$; therefore $A \times \varnothing = \varnothing$. Similarly $\varnothing \times A = \varnothing$.

Theorem 1. Let A, B, and C be any three sets. Then

(a) $A \times (B \cap C) = (A \times B) \cap (A \times C)$.

(b) $A \times (B \cup C) = (A \times B) \cup (A \times C)$.

Proof.

(a) $(a, x) \in A \times (B \cap C)$

$\Leftrightarrow (a \in A) \wedge (x \in B \cap C)$	Def. 1
$\Leftrightarrow (a \in A) \wedge (x \in B \wedge x \in C)$	Def. $\cap$
$\Leftrightarrow (a \in A) \wedge (a \in A) \wedge (x \in B) \wedge (x \in C)$	Idemp,. Assoc. (Ch.1)
$\Leftrightarrow [(a \in A) \wedge (x \in B)] \wedge [(a \in A) \wedge (x \in C)]$	Com., Assoc. (Ch. 1)
$\Leftrightarrow [(a, x) \in A \times B)] \wedge [(a, x) \in A \times C]$	Def. 1
$\Leftrightarrow (a, x) \in (A \times B) \cap (A \times C)$	Def. $\cap$

Hence, by Definition 1 of Chapter 2, we have proved that

$$A \times (B \cap C) = (A \times B) \cap (A \times C)$$

Intuitively, this equality indicates: *The Cartesian product distributes over intersection.*

We leave the proof of part (b) to the reader as an exercise.

Theorem 2. Let A, B, C be sets. Then

$$A \times (B - C) = (A \times B) - (A \times C)$$

That is, the Cartesian product distributes over complementation.

Proof.

$(a, x) \in A \times (B - C)$

$\Leftrightarrow (a \in A) \wedge (x \in B - C)$	Def. 1
$\Leftrightarrow (a \in A) \wedge [(x \in B) \wedge (x \notin C)]$	Def. 5 (Ch. 2)
$\Leftrightarrow (a \in A) \wedge (a \in A) \wedge (x \in B) \wedge (x \notin C)$	Idemp., Assoc. (Ch.1)
$\Leftrightarrow [(a \in A) \wedge (x \in B)] \wedge [(a \in A) \wedge (x \notin C)]$	Com., Assoc, (Ch. 1)
$\Leftrightarrow [(a, x) \in A \times B] \wedge [(a, x) \notin A \times C]$	Def. 1
$\Leftrightarrow (a, x) \in (A \times B) - (A \times C)$	Def. 5 (Ch. 2)

Thus, we have proved that

$$A \times (B - C) = (A \times B) - (A \times C)$$

EXERCISE 3.1

1. Describe each of the following sets geometrically by sketching a graph on the Cartesian plane.
   (a) $\{(x, y) \in \mathbf{R} \times \mathbf{R} \mid x = y\}$
   (b) $\{(x, y) \in \mathbf{R} \times \mathbf{R} \mid x > y\}$
   (c) $\{(x, y) \in \mathbf{R} \times \mathbf{R} \mid |x + y| \leq 1\}$
2. Under what conditions on the sets A and B will it be true that $A \times B = B \times A$?
3. Prove Theorem 1(b): $A \times (B \cup C) = (A \times B) \cup (A \times C)$.
4. Prove that $A \times B = \varnothing \Leftrightarrow A = \varnothing \vee B = \varnothing$.
5. Prove that, if A, B, and C are sets and $A \subseteq B$, then $A \times C \subseteq B \times C$.
6. If the set A has m elements and if the set B has n elements, how many elements (ordered pairs) does $A \times B$ have?
7. The Cartesian product $A \times A$ has nine elements among which are found $(-1, 0)$ and $(0, 1)$. Find the remaining elements and the set A.
8. Prove or disprove (by giving a counterexample) each of the following statements.
   (a) $A \times B \subseteq C \times D$ if and only if $A \subseteq C$ and $B \subseteq D$.
   (b) The power set $\mathscr{P}(A \times B)$ of $A \times B$ is the Cartesian product $\mathscr{P}(A) \times \mathscr{P}(B)$ of the power sets $\mathscr{P}(A)$ and $\mathscr{P}(B)$.
9. Prove that, if A, B, C, and D are any four sets, then
$$(A \times C) \cap (B \times D) = (A \cap B) \times (C \cap D).$$
10. Let $A_1, A_2, \ldots, A_n$ be sets. Can you generalize Definition 1 to the Cartesian product $A_1 \times A_2 \times A_3$ of three sets ? Can you generalize this further to the Cartesian product $A_1 \times A_2 \times \cdots \times A_n$ of n sets?
11. Prove that if $A \times A = B \times B$ then $A = B$.
12. Prove that if $A \times C = B \times C$ and $C \neq \varnothing$, then $A = B$.
13. Prove or disprove that $A \cup (B \times C) = (A \cup B) \times (A \cup C)$.
14. Prove that $(A \times B) \cap (C \times D) = (A \cap C) \times (B \cap D)$.
15. Is it true that $(A \times B) \cup (C \times D) = (A \cup C) \times (B \cup D)$?
16. Prove that if $A \cap B = \varnothing$ then $(A \times C) \cap (B \times D) = \varnothing$ for any sets C and D.
17. Let A, B, C, and D be nonempty sets. Prove that $A \times B = C \times D$ if and only if $A = C$ and $B = D$.
18. Prove that
$$(A \times B) - (C \times C) = (A - C) \times B \cup A \times (B - C).$$
19. Prove that
$$(A \times A) - (B \times C) = (A - B) \times A \cup A \times (A - C).$$
20. Prove that
$$(A \times B) - (C \times D) = (A - C) \times B \cup A \times (B - D).$$
21. Define the ordered pair (x, y) to be the set $\{\{x\}, \{x, y\}\}$. Use this definition to prove that $(a, b) = (c, d)$ if and only if $a = c$ and $b = d$.

2. RELATIONS

Given sets A and B, not necessarily distinct, when we say that an element a of A is related to another element b of B by a relation $\mathscr{R}$ we are making a statement about the ordered pair (a, b) in the Cartesian product $A \times B$. Therefore, a mathematical definition of a relation can be precisely given in terms of ordered pairs in the Cartesian product of sets.

Definition 2. A *relation* $\mathscr{R}$ *from* A to B is a subset of the Cartesian product $A \times B$. It is customary to write $a \mathscr{R} b$ for $(a, b) \in \mathscr{R}$. The symbol $a \mathscr{R} b$ is read "a is $\mathscr{R}$-related to b."

Often A and B are the same set, say X. In that case, we shall say that $\mathscr{R}$ is a relation *in* X instead of "from X to X." For example, in a community X,[2] to say that a (for Albert) is the husband of b (for Bonita), is to consider Albert and Bonita as an (ordered) pair (a, b) in the relation $\mathscr{H}$ (of being the husband of . . .). The symbol $a\mathscr{H}b$ or $(a, b) \in \mathscr{H}$ may be read "a is the husband of b."

It is not necessary to put Bonita behind Albert in the ordered pair (a, b). We may say that Bonita is the wife of Albert, or that the ordered pair (b, a) is the relation $\mathscr{W}$ (of being the wife of . . .).The symbol $b \mathscr{W} a$ or $(b, a) \in \mathscr{W}$ may be read: "b is the wife of a." In this example, the relation $\mathscr{W}$ is called the inverse of the relation $\mathscr{W}$.

Definition 3. Let A and B be two sets, not necessarily distinct, and let $\mathscr{R}$ be a relation from A to B. Then the *inverse* $\mathscr{R}^{-1}$ of the relation $\mathscr{R}$ is the relation from B to A such that $b \mathscr{R}^{-1} a$ if and only if $a \mathscr{R} b$. That is,

$$\mathscr{R}^{-1} = \{(b, a) \mid (a, b) \in \mathscr{R}\}$$

EXAMPLE 3. (a) Let $A = \{a, b\}$, $B = \{x, y, z\}$, and let $\mathscr{R} \subseteq A \times B$ be given by $\mathscr{R} = \{(a, x), (b, y)\}$. Then $\mathscr{R}^{-1} = \{(x, a), (y, a)\} \subseteq B \times A$.

(b) Let

$$\mathscr{R} = \{(x, y) \in \mathbf{N} \times \mathbf{N} \mid x \text{ divides } y\}$$

Then

$$\mathscr{R}^{-1} = \{(y, x) \in \mathbf{N} \times \mathbf{N} \mid y \text{ is a multiple of } x\}$$

Let $\mathscr{R}$ be a relation from A to B. The *domain* of the relation $\mathscr{R}$, denoted by $\mathrm{Dom}(\mathscr{R})$, is the set of all those $a \in A$ such that $a\mathscr{R}b$ for some

[2] Hence, X is the set of all members of the community.

$b \in B$; and the *image* of $\mathscr{R}$, denoted by $\text{Im}(\mathscr{R})$, is the set of all those $b \in B$ such that $a\mathscr{R}b$ for some $a \in A$. In symbols,

$$\text{Dom}(\mathscr{R}) = \{a \in A \mid (a, b) \in \mathscr{R} \text{ for some } b \in B\}$$

and

$$\text{Im}(\mathscr{R}) = \{b \in B \mid (a, b) \in \mathscr{R} \text{ for some } a \in A\}$$

In the example of relations $\mathscr{H}$ (being the husband of . . .) and $\mathscr{W}$ (being the wife of . . .) in the community X, the domain of $\mathscr{H}$ is the set of all men in X who are married, and the image of $\mathscr{H}$ is the set of all women in X who are married, whereas, the domain of $\mathscr{W}$ is the set of all wives in X, and the image of $\mathscr{W}$ is the set of all the husbands in X. That is,

$$\text{Dom}(\mathscr{W}) = \text{Im}(\mathscr{H})$$

and

$$\text{Im}(\mathscr{W}) = \text{Dom}(\mathscr{H})$$

Can you make a general conclusion? (See Problem 3 at the end of this section.)

EXAMPLE 4. In Example 3(a), $\text{Dom}(\mathscr{R}) = \{a, b\}$ and $\text{Im}(\mathscr{R}) = \{x, y\}$. In Example 3(b), $\text{Dom}(\mathscr{R}) = \mathbf{N} = \text{Im}(\mathscr{R})$.

Definition 4. Let $\mathscr{R}$ be a relation in a set X. Then we say that

(a) $\mathscr{R}$ is *reflexive* if and only if $\forall x \in X, x\mathscr{R}x$.
(b) $\mathscr{R}$ is *symmetric* if and only if $x\mathscr{R}y \Rightarrow y\mathscr{R}x$.
(c) $\mathscr{R}$ is *transitive* if and only if $x\mathscr{R}y \wedge y\mathscr{R}z \Rightarrow x\mathscr{R}z$.
(d) $\mathscr{R}$ is an *equivalence* relation if and only if $\mathscr{R}$ is reflexive, symmetric, and transitive.

The equals relation $=$, on the set $\mathbf{R}$ of real numbers is clearly an equivalence relation. Let X be a set of colored balls and let any two balls a and b be related by $\mathscr{R}$ if and only if a and b have the same color. Then the relation $\mathscr{R}$ is an equivalence relation.

Equivalence relations are particularly important in modern mathematics. For instance, factor groups in algebra, quotient spaces in topology, and modular systems in number theory all involve certain kinds of equivalence relations.

Given a nonempty set X, there always exist at least two equivalence

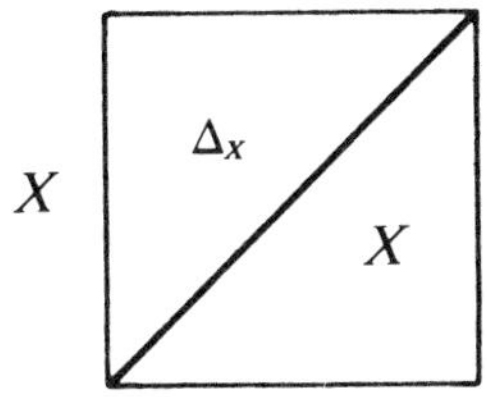

Figure 8.

relations in X; one of these is the *diagonal relation* Δ_X (also called the *identity relation*) defined by

$$\Delta_X = \{(x, x) \mid x \in X\}$$

which relates every element with itself. Pictorially, if X is represented as a line interval, then $X \times X$ is a square and Δ_X is the "main" diagonal of the square.

There is, at the other extreme, always another equivalence relation $\mathscr{R} = X \times X$ on[3] X. The relation Δ_X is the smallest of all the equivalence relations among the subsets of $X \times X$ that can be defined on X, whereas $X \times X$ is the largest.

EXAMPLE 5. Let m be an arbitary fixed positive integer. The *congruence relation* $\equiv$ modulo m on the set $\mathbf{Z}$ of integers is defined by $x \equiv y$(modulo m) if and only if $x - y = km$ for some $k \in \mathrm{Z}$. The congruence relation is an equivalence relation on $\mathbf{Z}$.

Proof.

(a) For each $x \in \mathbf{Z}$, since $x - x = 0 \cdot m$, we have $x \equiv x(\bmod m)$. Hence it is reflexive.

(b) If $x \equiv y(\bmod m)$, then $x - y = km$ for some $k \in \mathbf{Z}$. Consequently, $y - x \equiv (-k)m$ and $-k \in \mathbf{Z}$, or $y \equiv x(\bmod m)$. Hence it is symmetric.

(c) If $x \equiv y(\bmod m)$ and $y \equiv z(\bmod m)$, then $x - y = k_1 m$ and $y - z = k_2 m$ for some k_1 and k_2 in $\mathbf{Z}$. Hence $x - z = (x - y) + (y - z) = (k_1 + k_2)m$ and $k_1 + k_2 \in \mathbf{Z}$, which shows that $x \equiv z(\bmod m)$. Hence it is transitive.

---

[3] When the domain of a relation $\mathscr{R}$ is X itself, most mathematicians prefer to say "relation *on X*" instead of "relation *$\mathscr{R}$ in X*."

Therefore, we have proved that the congruence relation (modulo m) is an equivalence relation on $\mathbf{Z}$.

As a special case for Example 5, let $m = 2$. Then $x \equiv y \pmod 2$ if and only if $x - y$ is an even integer. Consequently, $x \equiv y \pmod 2$ if and only if either both x and y are even or both x and y are odd.

EXERCISE 3.2

1. Let $\mathcal{R}$ be a relation from A to B. Prove that $(\mathcal{R}^{-1})^{-1} = \mathcal{R}$.
2. Let $A = \{a, b, c\}$ and let $\mathcal{R} = \{(a, c), (c, b), (a, b)\}$. Find the domain of $\mathcal{R}$ and the image of $\mathcal{R}$.
3. Let $\mathcal{R}$ be a relation from A to B. Prove that
   (a) $\text{Dom}(\mathcal{R}^{-1}) = \text{Im}(\mathcal{R})$
   (b) $\text{Im}(\mathcal{R}^{-1}) = \text{Dom}(\mathcal{R})$.
4. Let $A = \{a, b, c\}$ and let
   $$\mathcal{R} = \{(a, a), (b, b), (c, c), (a, b), (b, a), (c, a), (a, c)\}$$
   Prove that $\mathcal{R}$ is reflexive and symmetric, but not transitive.
5. Give an example of a relation that is reflexive and transitive, but not symmetric.
6. Give an example of a relation that is symmetric and transitive, but not reflexive.
7. Let $\mathcal{R}$ be a relation in a set X. Prove that
   (a) $\mathcal{R}$ is reflexive if and only if $\mathcal{R} \supseteq \Delta_X$
   (b) $\mathcal{R}$ is symmetric if and only if $\mathcal{R} = \mathcal{R}^{-1}$
   (c) $\mathcal{R}$ is reflexive if and only if $\mathcal{R}^{-1}$ is reflexive
   (d) $\mathcal{R}$ is symmetric if and only if $\mathcal{R}^{-1}$ is symmetric
   (e) $\mathcal{R}$ is transitive if and only if $\mathcal{R}^{-1}$ is transitive
   (f) $\mathcal{R}$ is an equivalence relation if and only if $\mathcal{R}^{-1}$ is an equivalence relation.
8. How many relations are there on a set with n elements?
9. How many relations are there from a set with m elements to a set with n elements?
10. Let $\mathcal{R}$ be a relation from A to B and $D \subseteq A$. By the *restriction* of $\mathcal{R}$ to D we mean the relation $\mathcal{R} \mid D = \{(x, y) \in \mathcal{R} \mid x \in D\}$ from D to B. Prove that $\mathcal{R} \mid D = \mathcal{R} \cap (D \times \text{Im}(\mathcal{R}))$.
11. Prove that if $\mathcal{R}$ is a relation from A to B and if D and E are subsets of A, then
    (a) $\mathcal{R} \mid (D \cup E) = (\mathcal{R} \mid D) \cup (\mathcal{R} \mid E)$
    (b) $\mathcal{R} \mid (D \cap E) = (\mathcal{R} \mid D) \cap (\mathcal{R} \mid E)$.
12. Let $\mathcal{R}$ be a relation from A to B and let X be any subset of A. Define $\mathcal{R}(X)$, called the $\mathcal{R}$*-image* of X, by

$$\mathscr{R}(X) = \{y \in B \mid (x, y) \in \mathscr{R} \text{ for some } x \in X\}.$$

Prove that if D and E are subsets of A, then
(a) $\mathscr{R}(D \cup E) = \mathscr{R}(D) \cup \mathscr{R}(E)$
(b) $\mathscr{R}(D \cap E) \subseteq \mathscr{R}(D) \cap \mathscr{R}(E)$.

13. Prove, under the assumptions of Problem 12, that
(a) $\text{Dom}(\mathscr{R}) = \mathscr{R}^{-1}(B)$
(b) $\text{Im}(\mathscr{R}) = \mathscr{R}(A)$.

14. If $\mathscr{R}$ and $\mathscr{S}$ are relations from A to B, then $\mathscr{R} \cup \mathscr{S}$ is a relation from A to B. Prove that
(a) $\text{Dom}(\mathscr{R} \cup \mathscr{S}) = \text{Dom}(\mathscr{R}) \cup \text{Dom}(\mathscr{S})$
(b) $\text{Im}(\mathscr{R} \cup \mathscr{S}) = \text{Im}(\mathscr{R}) \cup \text{Im}(\mathscr{S})$
(c) $(\mathscr{R} \cup \mathscr{S})(X) = \mathscr{R}(X) \cup \mathscr{S}(X)$, for all $X \subseteq A$.

15. Let $\mathscr{R}$ be a relation on X. Prove that $\mathscr{R} \cup \mathscr{R}^{-1}$ is a symmetric relation on X and that if $\mathscr{S}$ is a symmetric relation containing $\mathscr{R}$ then $\mathscr{S} \supseteq \mathscr{R} \cup \mathscr{R}^{-1}$. Thus, $\mathscr{R} \cup \mathscr{R}^{-1}$ is the smallest symmetric relation containing $\mathscr{R}$.

16. Let $\mathscr{R}$ be a relation on X. Prove that $\mathscr{R} \cap \mathscr{R}^{-1}$ is a symmetric relation of X and that if $\mathscr{S}$ is a symmetric relation such that $\mathscr{S} \subseteq \mathscr{R}$ then $\mathscr{S} \subseteq \mathscr{R} \cap \mathscr{R}^{-1}$. Thus, $\mathscr{R} \cap \mathscr{R}^{-1}$ is the largest symmetric relation contained in $\mathscr{R}$.

17. Prove that
(a) $(\mathscr{R} \cup \mathscr{S})^{-1} = \mathscr{R}^{-1} \cup \mathscr{S}^{-1}$
(b) $(\mathscr{R} \cap \mathscr{S})^{-1} = \mathscr{R}^{-1} \cap \mathscr{S}^{-1}$.

18. Let $X = \mathbf{Z} \times (\mathbf{Z} - \{0\})$. Define a relation $\sim$ on X by declaring that $(a, b) \sim (c, d)$ if and only if $ad = bc$. Prove that the relation $\sim$ is an equivalence relation.

3. PARTITIONS AND EQUIVALENCE RELATIONS

Definition 5. Let X be a nonempty set. By a *partition* $\mathfrak{P}$ of X we mean a set of nonempty subsets of X such that
(a) If $A, B \in \mathfrak{P}$ and $A \neq B$, then $A \cap B = \varnothing$.
(b) $\bigcup_{C \in \mathfrak{P}} C = X$.

Intuitively, a partition of X is a "cutting up" of X into (nonempty disjoint) pieces.

EXAMPLE 6. Let m be any fixed positive integer. For each integer j, $0 \leq j < m$, let $\mathbf{Z}_j = \{x \in \mathbf{Z} \mid x - j = km \text{ for some } k \in \mathbf{Z}\}$. Then the set

$$\{\mathbf{Z}_0, \mathbf{Z}_1, \mathbf{Z}_2, \ldots, \mathbf{Z}_{m-1}\}$$

forms a partition of $\mathbf{Z}$. In particular, let $m = 2$. Then the set of sets

$$\mathbf{Z}_0 = \{x \in \mathbf{Z} \mid x \text{ is even}\}$$

and

$$\mathbf{Z}_1 = \{x \in \mathbf{Z} \mid x - 1 \text{ is even}\} = \{x \in \mathbf{Z} \mid x \text{ is odd}\}$$

forms a partition of $\mathbf{Z}$. (See also Problem 8, Exercise 3.3)

There is a very close connection between the partition of a nonempty set and an equivalence relation on that set. In order to understand this connection, we shall first need the following definition.

Definition 6. Let $\mathscr{E}$ be an equivalence relation on a nonempty set X. For each $x \in X$, we define

$$x/\mathscr{E} = \{y \in X \mid y \mathscr{E} x\}$$

which is called the *equivalence class* determined by the element x. The set of all such equivalence classes in X is denoted by $X/\mathscr{E}$; that is, $X/\mathscr{E} = \{x/\mathscr{E} \mid x \in X\}$. The symbol $X/\mathscr{E}$ is read "X modulo $\mathscr{E}$," or simply "X mod $\mathscr{E}$."

Theorem 3. Let $\mathscr{E}$ be an equivalence relation on a nonempty set X. Then

(a) Each $x/\mathscr{E}$ is a nonempty subset of X.
(b) $x/\mathscr{E} \cap y/\mathscr{E} \neq \varnothing$ if and only if $x \mathscr{E} y$.
(c) $x \mathscr{E} y$ if and only if $x/\mathscr{E} = y/\mathscr{E}$.

Proof.

(a) Since $\mathscr{E}$ is reflexive, for each $x \in X$, we have $x \mathscr{E} x$. By Definition 6, $x \in x/\mathscr{E}$ and hence $x/\mathscr{E}$ is a nonempty subset of X.

(b) Since $\mathscr{E}$ is an equivalence relation and $X \neq \varnothing$, we have

$x/\mathscr{E} \cap y/\mathscr{E} \neq \varnothing \Leftrightarrow (\exists z)(z \in x/\mathscr{E} \wedge z \in y/\mathscr{E})$	
$\Leftrightarrow (z \mathscr{E} x) \wedge (z \mathscr{E} y)$	Def. 6
$\Leftrightarrow (x \mathscr{E} z) \wedge (z \mathscr{E} y)$	$\mathscr{E}$ is symmetric
$\Leftrightarrow x \mathscr{E} y$	$\mathscr{E}$ is transitive

(c) It follows immediately from (a) and (b) above that $x/\mathscr{E} = y/\mathscr{E} \Rightarrow x \mathscr{E} y$. We now need to prove that $x \mathscr{E} y \Rightarrow x/\mathscr{E} = y/\mathscr{E}$. Let $x \mathscr{E} y$. Then

$z \in x/\mathscr{E} \Rightarrow z \mathscr{E} x$	Def.6
$(z \mathscr{E} x) \wedge (x \mathscr{E} y) \Rightarrow z \mathscr{E} y$	$\mathscr{E}$ is transitive
$\Rightarrow z \in y/\mathscr{E}$	Def. 6

Since z is arbitrary, it follows that $x/\mathscr{E} \subseteq y/\mathscr{E}$. A similar argument gives $y/\mathscr{E} \subseteq x/\mathscr{E}$; hence $x/\mathscr{E} = y/\mathscr{E}$.

Theorem 4. Let $\mathscr{E}$ be an equivalence relation on a nonempty set X. Then $X/\mathscr{E}$ is a partition of X.

Proof. By Theorem 3(a) and Definition 6, $X/\mathscr{E} = \{x/\mathscr{E} \mid x \in X\}$ is a family of nonempty subsets of X. We next show that

$$x/\mathscr{E} \neq y/\mathscr{E} \Rightarrow (x/\mathscr{E}) \cap (y/\mathscr{E}) = \varnothing$$

by showing its contrapositive: $[x/\mathscr{E} \cap y/\mathscr{E} \neq \varnothing] \Rightarrow [x/\mathscr{E} = y/\mathscr{E}]$. The last assertion is a direct consequence of Theorem 3(b) and (c). Finally, we have to show that $\bigcup_{x \in X} x/\mathscr{E} = X$. This is also trivial, since each x in X belongs to $x/\mathscr{E}$. This completes the proof of the theorem.

We have seen, in Theorem 4, that an equivalence relation on the nonempty set X gives rise to a partition of X. We shall next show that the converse of Theorem 4 is true; that is, each partition of X gives rise to an equivalence relation on X.

Definition 7. Let $\mathscr{P}$ be a partition of a nonempty set X. We define a relation $X/\mathscr{P}$ on X by $x(X/\mathscr{P})y$ if and only if there exists a set $A \in \mathscr{P}$ such that x, $y \in A$.

Warning! The reader should read and compare Definitions 6 and 7 carefully to understand the delicate differences among these similar notations: $x/\mathscr{E}$, $X/\mathscr{E}$, and $X/\mathscr{P}$.

Theorem 5. Let $\mathscr{P}$ be a partition of a nonempty set X. Then the relation $X/\mathscr{P}$ is an equivalence relation on X, and the equivalence classes induced by the equivalence relation $X/\mathscr{P}$ are precisely the sets in $\mathscr{P}$. Symbolically, $X/(X/\mathscr{P}) = \mathscr{P}$.

Proof. Since every element x of X is contained in some $A \in \mathscr{P}$, $x(X/\mathscr{P})x$; that is, $X/\mathscr{P}$ is reflexive. The symmetry of $X/\mathscr{P}$ is a clear consequence of Definition 7. To show that the relation $X/\mathscr{P}$ is transitive, let x, y, and z be three elements of X satisfying

$$x(X/\mathscr{P})y \qquad \text{and} \qquad y(X/\mathscr{P})z$$

Then, by Definition 7, there exist A and B in $\mathscr{P}$ such that x, $y \in A$ and y, $z \in B$. Consequently, $y \in A \cap B \neq \varnothing$. It follows, by the definition of a partition, that $A = B$. Hence, x, $z \in A$ and hence $x(X/\mathscr{P})z$. Thus, $X/\mathscr{P}$ is an equivalence relation on X.

To show the remainder of the theorem, let x be an arbitrary element of X. There exists one and only one set A in $\mathcal{T}$ such that $x \in A$. (Why?) Consequently, by Definition 6, we have

$$x/(X/\mathcal{T}) = A$$

We have just proved that each equivalence class modulo $X/\mathcal{T}$ is a set in the family $\mathcal{T}$. Conversely, let A be any set in the partition $\mathcal{T}$. Since $A \neq \emptyset$, there exists an element x in X that belongs to A. By our previous argument, $x/(X/\mathcal{T})\, A$. This proves that $X/(X/\mathcal{T}) = \mathcal{T}$. The proof of the theorem is now complete.

Any equivalence relation $\mathcal{E}$ on the nonempty set X gives rise to a partition $X/\mathcal{E}$ (Theorem 4); this partition in turn determines an equivalence relation $X/(X/\mathcal{E})$ (Theorem 5). The crucial fact is that $X/(X/\mathcal{E}) = \mathcal{E}$ (see Problem 6). This together with $X/(X/\mathcal{T}) = \mathcal{T}$ establishes the intimate connection between equivalence relations and partitions.

Let us illustrate Theorem 5 by a concrete example. Let $\mathbf{Z}_0$ and $\mathbf{Z}_1$ be the set of even integers and the set of odd integers, respectively. Then $\mathcal{T} = \{\mathbf{Z}_0, \mathbf{Z}_1\}$ forms a partition of the set $\mathbf{Z}$ of integers. By definition of the relation $\mathbf{Z}/\mathcal{T}$, we have $a(\mathbf{Z}/\mathcal{T})b$ if and only if either $a, b \in \mathbf{Z}_0$ or $a, b \in \mathbf{Z}_1$. That is $a(\mathbf{Z}/\mathcal{T})b$ if and only if either both a and b are even of both a and b are odd. It is easy to verify that this relation $\mathbf{Z}/\mathcal{T}$ is indeed an equivalence relation. In effect, $a(\mathbf{Z}/\mathcal{T})b$ if and only if $a \equiv b \pmod 2$. Therefore, the relation $\mathbf{Z}/\mathcal{T}$ is really the familiar congruence relation $\equiv \pmod 2$. [See Example 5.]

Conversely, given the set $\mathbf{Z}$ together with the equivalence relation $\mathcal{E}$ such that $x\,\mathcal{E}\,y$ if and only if $x \equiv y \pmod 2$, then

$$a/\mathcal{E} = \{x \in \mathbf{Z} \mid x \equiv a \pmod 2\} = \begin{cases} \mathbf{Z}_0 & \text{if } a \text{ is even} \\ \mathbf{Z}_1 & \text{if } a \text{ is odd} \end{cases}$$

Therefore, $\mathbf{Z}/\mathcal{E} = \{\mathbf{Z}_0, \mathbf{Z}_1\}$, which is clearly a partition of $\mathbf{Z}$.

EXERCISE 3.3

1. Let $X = \{a, b, c, d, e\}$, $A = \{a, b\}$ and $B = \{c, d, e\}$
   (a) Verify that $\{A, B\}$ is a partition of X.
   (b) Identify the relation $X/\{A, B\}$ induced by the partition $\{A, B\}$.
2. Let $X = \{a, b, c, d\}$ and $\mathcal{R} = \{(a, b), (b, a), (a, a), (b, b), (c, d), (d, c), (c, c), (d, d)\}$.
   (a) Verify that $\mathcal{R}$ is an equivalence relation on X.
   (b) Find the partition $X/\mathcal{R}$ induced by $\mathcal{R}$.
3. Let $\mathcal{T}$ be the partition $X/\mathcal{R}$ of Problem 2, above. Find the relation $X/\mathcal{T}$ on X induced by the partition $\mathcal{T}$.
4. Let $\mathcal{T}$ be a partition of the nonempty set X. Prove that the

equivalence relation $X/\mathcal{T} = \bigcup_{A \in \mathcal{T}} A \times A$.

5. In Problem 1, how many elements are contained in $A \times A \cup B \times B$? How many ordered pairs in $X/\{A, B\}$?
6. In Problem 1, let X be a finite set and let

$$\mathcal{T} = \{A_1, A_2, \ldots, A_k\}$$

where the set A_j contains n_j elements for $j = 1, 2, \ldots, k$. Prove that the number of ordered pairs in the equivalence relation $X/\mathcal{T}$ is exactly $n_1^2 + n_2^2 + \ldots + n_k^2$.

7. Let $X = \{a, b, c, d, e\}$ and let $\mathcal{T} = \{\{a, b\}, \{c\}, \{d, e\}\}$.
   (a) Show that $\mathcal{T}$ is a partition of X.
   (b) Find the equivalence relation $X/\mathcal{T}$ on X explicity as a set of ordered pairs.
   (c) Denote $\mathcal{E} = X/\mathcal{T}$ and find $a/\mathcal{E}$, $b/\mathcal{E}$, $c/\mathcal{E}$, $d/\mathcal{E}$, and $e/\mathcal{E}$ explicitly.
8. Verify Example 6 for $m = 3$.
9. Let X be the set $\mathbf{Z}$ of integers and let $\mathcal{E}$ be a relation on X defined by $x \mathcal{E} y$ if and only if $x - y = 5k$ for some integer k.
   (a) Prove that the relation $\mathcal{E}$ is an equivalence relation on X.
   (b) Find the partition $X/\mathcal{E}$ of X.
   (c) Verify that the equivalence relation $X/(X/\mathcal{E})$ is indeed the equivalence relation $\mathcal{E}$.
10. Let $\{A_1, A_2, \ldots, A_m\}$ be a partition of set A, and let $\{B_1, B_2, \ldots, B_n\}$ be a partition of a set B. Prove that $\{A_i \times B_j \mid i = 1, 2, \ldots, m; j = 1, 2, \ldots, n\}$ is a partition of $A \times B$.
11. Let $\mathcal{E}$ be an equivalence relation on the nonempty set X. Prove that $X/(X/\mathcal{E}) = \mathcal{E}$.

4. FUNCTIONS

Unquestionably, the concept of a function is one of the most basic ideas in every branch of mathematics. The reader may have learned the following definition: a function is a *rule* of correspondence that assigns to each element x of a certain set (called the *domain* of the function) one and only one element y in another set (called the *range* of the function). This definition is cloudy. What is meant precisely by a "rule"? In order to avoid ambiguities, mathematicians have devised a precise definition of a function using the language of sets.

Definition 8. Let X and Y be sets. A *function from X to Y* is a triple (f, X, Y), where f is a relation from X to Y satisfying
(a) $\text{Dom}(f) = X$.
(b) If $(x, y) \in f$ and $(x, z) \in f$, then $y = z$.

Let (f, X, Y) be a function from X to Y. In what follows, we shall adhere to the custom of writing $f: X \rightarrow Y$ instead of (f, X, Y), and $y = f(x)$ instead of $(x, y) \in f$. The reason "$y = f(x)$" is a meaningful substitute for "$(x, y) \in f$" is

Every element $x \in X$ has a uniquely determined $y \in Y$ such that $(x, y) \in f$.

To see that this assertion is true, let $x \in X$. Then by condition (a) of Definition 8, there exists an element $y \in Y$ such that $(x, y) \in f$; if there is another element $z \in Y$ with $(x, z) \in f$, then according to condition (b), $z = y$. This shows that y is uniquely determined by $x \in X$.

Let $f: X \rightarrow Y$ be a function. If $y = f(x)$, we say that y is the image of x under f and that x is a *preimage* of y under f. The reader may picture this as illustrated in Figures 9 and 10. We call the set Y, in $f: X \rightarrow Y$, the *range* of the function. The reader should notice that the range of a function need not be the same as the image of the function[4] (see Example 7, below) We call the reader's attention to the fact that some authors use the word "range" to mean "image", but for a technical reason, which will be apparent in Section 6, we make a distinction between "image" and "range" of a function. In general, the image of a function is a subset of the range of that function.

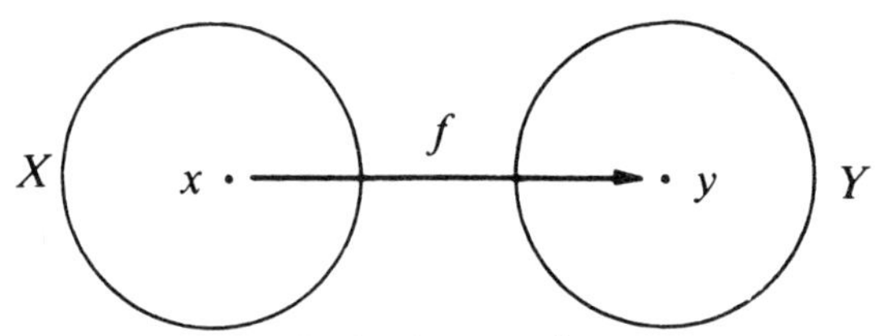

y is the image of x

Figure 9.

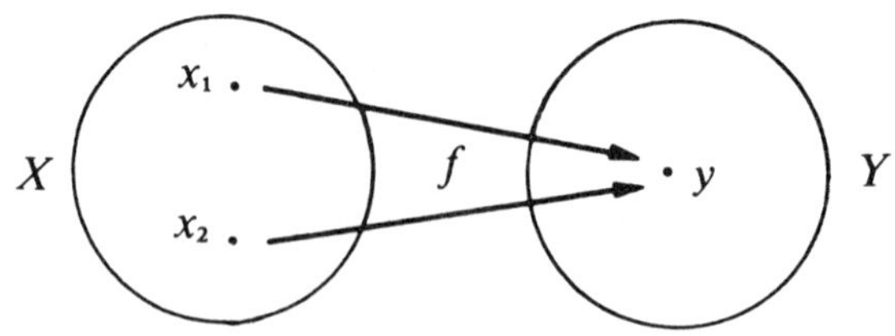

x_1 and x_2 are preimages of y

Figure 10.

[4] The *image* of the function $f: X \rightarrow Y$ is the image, $\mathrm{Im}(f)$, of the relation f. Consequently, $\mathrm{Im}(f) = \{f(x) \mid x \in X\}$.

EXAMPLE 7. Let $f: \mathbf{R} \to \mathbf{R}$ be defined by $f(x) = [x]$ for all x in $\mathbf{R}$, where $[x]$ denotes the greatest integer $\leq x$, e.g., $[\sqrt{2}] = 1, [-\frac{1}{2}] = -1$. Here the range of f is $\mathbf{R}$, whereas the image of f is $\mathbf{Z}$, a proper subset of $\mathbf{R}$.

It is possible to alter the range of a function without otherwise changing the function. For instance, for the same relation f as in Example 7 above, $f: \mathbf{R} \to \mathbf{Q}$ and $f: \mathbf{R} \to \mathbf{Z}$ are functions, because Definition 8 is satisfied. In general, we have the following theorem.

Theorem 6. Let $f: X \to Y$ be a function and let W be a set containing the image of f. Then $f: X \to W$ is a function.

Proof. We first prove that f is a relation from X to W:

$(x, y) \in f \Rightarrow x \in X \wedge y \in (\operatorname{Im} f)$	Def. of Im
$\Rightarrow x \in X \wedge y \in W$	$\operatorname{Im}(f) \subseteq W$
$\Rightarrow (x, y) \in X \times W$	Def. 1

This proves that $f \subseteq X \times W$; in other words, f is a relation from X to W. Now since $f: X \to Y$ is a function, $\operatorname{Dom}(f) = X$ and condition (b) of Definition 8 is satisfied. Therefore, $f: X \to W$ is a function.

Theorem 7. Let $f: X \to Y$ and $g: X \to Y$ be functions. Then $f = g$ if and only if $f(x) = g(x), \ \forall x \in X$.

Proof. (1) Suppose that $f = g$ and that x is an arbitrary element in X. Then,

$y = f(x) \Leftrightarrow (x, y) \in f$	Notation
$\Leftrightarrow (x, y) \in g$	$f = g$
$\Leftrightarrow g(x) = y$	Notation

Hence, $f(x) = g(x)$.

(2) Suppose that $f(x) = g(x), \ \forall x \in X$. Then

$(x, y) \in f \Leftrightarrow y = f(x)$	Notation
$\Leftrightarrow y = g(x)$	$f(x) = g(x)$
$\Leftrightarrow (x, y) \in g$	Notation

This proves that $f = g$.

If the domain and the range of a function are subsets of the set of real numbers, then, as in analytic geometry, the graph of the function may be sketched on the Cartesian plane. For example, the function in Example 7 has the following graph.

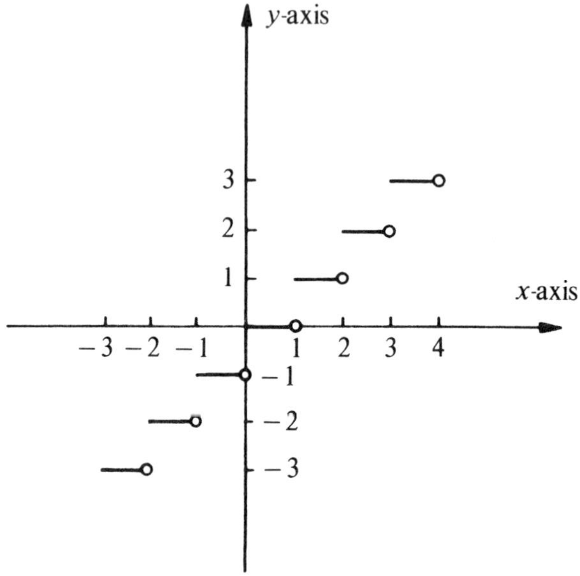

Figure 11.

EXAMPLE 8. Let A be a subset of a nonempty set X. Then the relation

$$\{(x, y) \in X \times \{0, 1\} \mid y = 1 \text{ if } x \in A \text{ and } y = 0 \text{ if } x \notin A\}$$

gives rise to a function from X to $\{0, 1\}$, known as the *characteristic function* of A in X. This function is usually denoted by the Greek letter chi with a subscript A, χ_A. That is,

$$\chi_A : X \to \{0, 1\}$$

is defined by

$$\chi_A(x) = \begin{cases} 1 & \text{if } x \in A \\ 0 & \text{if } x \in X - A \end{cases}$$

Although a function is, by definition, written (f, X, Y) or $f: X \to Y$, it is often a nuisance to have to write the domain and the range of the function explicitly when they are implicitly clear from the context. Therefore, we *shall denote a function by f when the domain and the range of f are clearly understood, without explicitly giving the domain and the range of f.*

EXAMPLE 9. Let X be a set. The diagonal relation Δ_X on X defined on page 56 is a function from X to X. When we wish to stress that the relation Δ_X is a function, we use the alternative notation $I_X : X \to X$, where $I_X(x) = x$ for all x in X. The function I_X is called the *identity function* on X.

EXAMPLE 10. Let X and Y be two nonempty sets and let b be a fixed element of Y. The relation

$$C_b = \{(x, b) \mid x \in X\}$$

gives rise to the function $C_b : X \to Y$ given by $C_b(x) = b$ for all x in X. The function C_b is called a *constant function*.

In calculus, we have often seen a function defined by two (or more) rules of correspondence: for example, $h : \mathbf{R} \to \mathbf{R}$ defined by

$$h(x) = \begin{cases} 1 - 2x, & \text{if } x \leq 0 \\ x^2 + 1, & \text{if } x \geq 0 \end{cases}$$

This function may be considered as the union of the following two functions:
(1) $f: (-\infty, 0] \to R$ defined by $f(x) = 1 - 2x$, $\forall x \in (-\infty, 0]$
(2) $g: [0, \infty) \to R$ defined by $g(x) = x^2 + 1$, $\forall x \in [0, \infty)$
The reader should notice that here $\text{Dom}(f) \cap \text{Dom}(g) = \{0\}$ and that $f(0) = g(0)$.

The last example motivates the following general theorem.

Theorem 8. Let $f: A \to C$ and $g: B \to D$ be two functions such that $f(x) = g(x)$, $\forall x \in A \cap B$. Then the *union* h of f and g defines the function

$$h = f \cup g : A \cup B \to C \cup D$$

where

$$h(x) = \begin{cases} f(x), & \text{if } x \in A \\ g(x), & \text{if } x \in B \end{cases}$$

Proof. Since f and g are relations, $f \subseteq A \times C$ and $g \subseteq B \times D$, and we have

$$h = f \cup g \subseteq (A \times C) \cup (B \times D)$$

$$\subseteq (A \cup B) \times (C \cup D)$$

because both $A \times C$ and $B \times D$ are subsets of $(A \cup B) \times (C \cup D)$. Thus, h is a relation from $A \cup B$ to $C \cup D$. We leave it to the reader to verify that

$$\text{Dom}(h) = \text{Dom}(f) \cup \text{Dom}(g)$$

$$= A \cup B$$

This shows that the relation h satisfies Definition 8(a).

For each element $x \in A \cup B$, we may consider the following three cases: (1) $x \in A - B$, (2) $x \in B - A$, and (3) $x \in A \cap B$. Since $f: A \to C$ and $g: B \to D$ satisfy Definition 8(b) and $f(x) = g(x) \forall x \in A \cap B$, we have that $h(x)$ is uniquely defined in each of the three cases. Thus the relation h satisfies Definition 8(b) as well. Hence, $h: A \cup B \to C \cup D$ is indeed a function.

EXERCISE 3.4

1. Test whether or not each of the following diagrams defines a function from $X = \{x, y, z\}$ to $Y = \{u, v, w\}$.

(a)

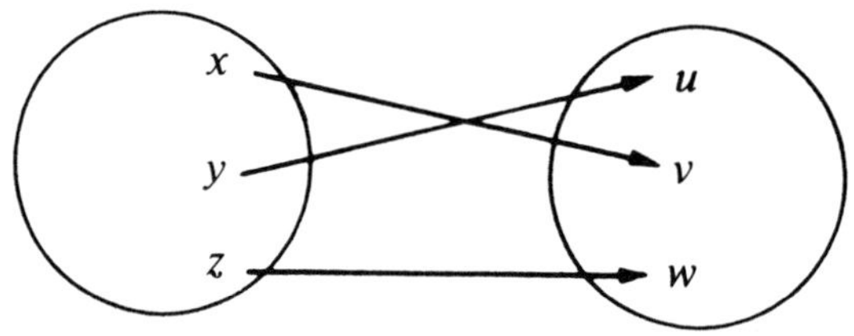

(b)

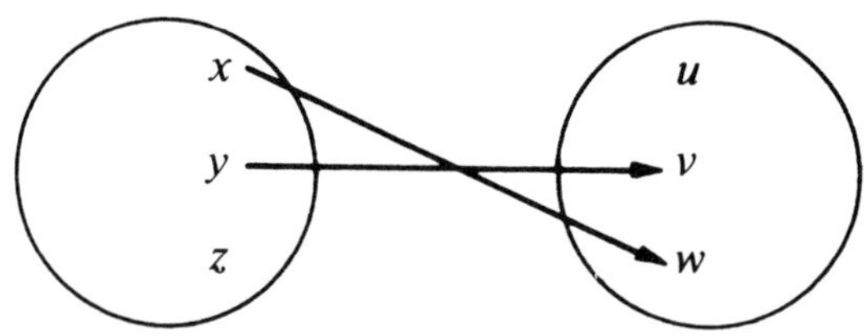

(c)

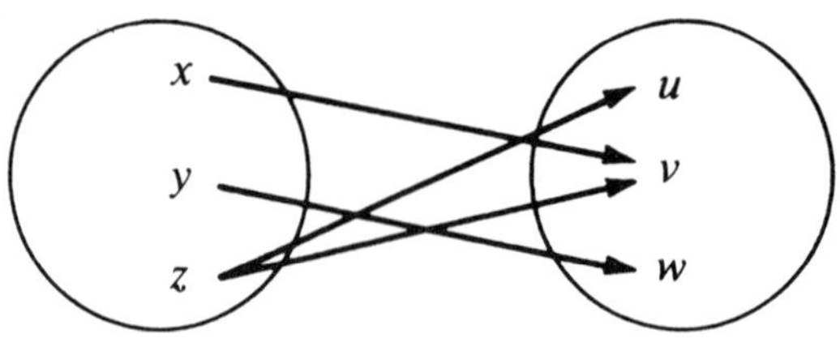

2 . Let $f: \mathbf{R} \to \mathbf{R}$ be the function given by

$$f(x) = \begin{cases} 5 & \text{if } x \text{ is rational} \\ -3 & \text{if } x \text{ is irrational} \end{cases}$$

Find $f(1/3)$, $f(7)$, and $f(1.323232\cdots)$.

3. Let the function $f: \mathbf{R} \to \mathbf{R}$ be given by

$$f(x) = \begin{cases} 4x + 3 & \text{if } x > 5 \\ x^2 - 2 & \text{if } -6 \leq x \leq 5 \\ 4 - 5x & \text{if } x < -6 \end{cases}$$

Find $f(-7)$, $f(3)$, and $f(6)$.

4. Let the function $f: X \to Y$ be defined by the diagram

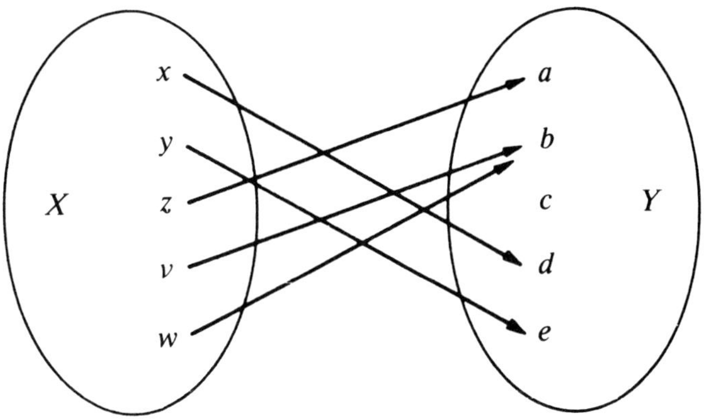

What is the image of this function?

5. Let the function $f: X \to \mathbf{R}$ be defined by $X = \{-2, -1, 0, 1, 2\}$ and $f(x) = x^2 - 3$ for all $x \in X$. Find the image of the function f.
6. Each of the following expressions defines a function from $\mathbf{R}$ to $\mathbf{R}$. Find the image of each function.
   (a) $f(x) = 2x^2 + 5$
   (b) $g(x) = \cos x$
   (c) $h(x) = x^3 - 1$
7. Let $X \subseteq Y$ and $f = \{(x, x) \mid x \in X\}$. Prove that $f: X \to Y$ is a function. [*Remark.* This function is called an *inclusion function* which may be denoted be $f: X \subseteq Y$.]

8. Let $X = \{x, y, z\}$ and $Y = \{1, 2, 3\}$. Which of the following constitute functions from X to Y? If they do not, give the reason.
   (a) $f = \{(x, 1), (y, 2), (z, 3)\}$
   (b) $g = \{(x, 2), (y, 3), (z, 2)\}$
   (c) $h = \{(x, 2), (y, 1)\}$
   (d) $i = \{(x, 1), (x, 2), (y, 1), (z, 3)\}$
9. In Problem 8, above,
   (a) Find the inverse relations f^{-1}, g^{-1}, h^{-1} and i^{-1};
   (b) Which of the inverse relations in (a), above are functions from Y to X, and which are not? Why?
10. Is f^{-1} of Problem 4 a function from Y to X?
11. If $X = \{x, y, z\}$ and $Y = \{1, 2\}$, how many functions from X to Y exist? More generally, if the set X has m elements and if Y has n elements, how many functions from X to Y exist?
12. How many of the functions in Problem 11 are constant functions?
13. Let $f : X \rightarrow Y$ be a function and let A be a nonempty subset of X. Prove that the restriction $f|A : A \rightarrow Y$ of $f : X \rightarrow Y$ is a function. [For the definition of "restriction" see Problem 10, Exercise 3.2.]
14. Let $f : X \rightarrow Y$ be a function. Prove that every subset g of f gives rise to a function.
15. Let $f : X \rightarrow X$ *be a function from* X to X that is also a reflexive relation on X. Prove that f must be the identity function $I_X : X \rightarrow X$.
16. Let X be the unit interval [0, 1]. Find a function $f : X \rightarrow X$ that is a symmetric relation on X.
17. Let $f : X \rightarrow Y$ and $g : X \rightarrow Y$ be two functions with the same domain and the same range. Prove that if $f \subseteq g$ then $f = g$.

5. IMAGES AND INVERSE IMAGES OF SETS

Recall that if $f : X \rightarrow Y$ is a function and if x and y are elements of X and Y, respectively, such that $y = f(x)$, then y is the image of x, and x is a preimage of y. This concept can be extended naturally from elements to subsets as follows:

Definition 9. Let $f : X \rightarrow Y$ be a function, and let A and B be subsets of X and Y, respectively.
   (a) The *image* of A under f, which we denote $f(A)$, is the set of all images $f(x)$ such that $x \in A$.
   (b) The *inverse image* of B under f, which we denote $f^{-1}(B)$, is the set of all preimages of y in B.

Using set builder notation, we have the following expressions:

$$f(A) = \{f(x) \mid x \in A\}$$
$$f^{-1}(B) = \{x \mid f(x) \in B\}$$

Theorem 9. Let $f: X \to Y$ be a function. Then

(a) $f(\varnothing) = \varnothing$.

(b) $f(\{x\}) = \{f(x)\}\ \forall x \in X$.

(c) If $A \subseteq B \subseteq X$, then $f(A) \subseteq f(B)$.

(d) If $C \subseteq D \subseteq Y$, then $f^{-1}(C) \subseteq f^{-1}(D)$.

Theorem 9 follows easily from Definition 9; therefore the proof is left to the reader.

Theorem 10. Let $f: X \to Y$ be a function and let $\{A_\gamma \mid \gamma \in \Gamma\}$ be a family of subsets of X. Then

(a) $f(\bigcup_{\gamma \in \Gamma} A_\gamma) = \bigcup_{\gamma \in \Gamma} f(A_\gamma)$.

(b) $f(\bigcap_{\gamma \in \Gamma} A_\gamma) \subseteq \bigcap_{\gamma \in \Gamma} f(A_\gamma)$.

Proof. (a) By repeated use of Definition 9 and Definition 6 of Chapter 2, we have

$$\begin{aligned} y \in f\left(\bigcup_{\gamma \in \Gamma} A_\gamma\right) &\Leftrightarrow y = f(x) \quad \text{for some} \quad x \in \bigcup_{\gamma \in \Gamma} A_\gamma \\ &\Leftrightarrow y = f(x) \quad \text{for some} \quad x \in A_\gamma, \quad \text{for some} \quad \gamma \in \Gamma \\ &\Leftrightarrow y \in f(A\gamma), \quad \text{for some} \quad \gamma \in \Gamma \\ &\Leftrightarrow y \in \bigcup_{\gamma \in \Gamma} f(A_\gamma) \end{aligned}$$

Therefore, $f(\bigcup_{\gamma \in \Gamma} A_\gamma) = \bigcup_{\gamma \in \Gamma} f(A_\gamma)$.

(b) Since $\bigcap_{\gamma \in \Gamma} A_\gamma \subseteq A_\gamma$ for every $\gamma \in \Gamma$, by Theorem 9(c), we have $f(\bigcap_{\gamma \in \Gamma} A_\gamma) \subseteq f(A_\gamma)$, for every $\gamma \in \Gamma$. It follows from Definition 7 of Chapter 2 that $f(\bigcap_{\gamma \in \Gamma} A_\gamma) \subseteq \bigcap_{\gamma \in \Gamma} f(A_\gamma)$.

The inclusion symbol $\subseteq$ in Theorem 10(b) may not be replaced by an equals sign, as the next example shows.

EXAMPLE 11. Let $X = \{a, b\}$, $Y = \{c\}$, $\Gamma = \{1, 2\}$, $A_1 = \{a\}$, $A_2 = \{b\}$, and let $f: X \to Y$ be the constant function, $f(a) = f(b) = c$. Then $f(A_1 \cap A_2) = f(\varnothing) = \varnothing$, whereas $f(A_1) \cap f(A_2) = \{c\}$. This shows that $f(\bigcap_{\gamma \in \Gamma} A_\gamma) = \bigcap_{\gamma \in \Gamma} f(A_\gamma)$ is not true in general.

Theorem 11. Let $f: X \to Y$ be a function and let $\{B_\gamma \mid \gamma \in \Gamma\}$ be a family of subsets of Y. Then
(a) $f^{-1}(\bigcup_{\gamma\in\Gamma} B_\gamma) = \bigcup_{\gamma\in\Gamma} f^{-1}(B_\gamma)$,
(b) $f^{-1}(\bigcap_{\gamma\in\Gamma} B_\gamma) = \bigcap_{\gamma\in\Gamma} f^{-1}(B_\gamma)$.

Proof. (a) By repeated applications of Definition 9, and Definition 6 of Chapter 2, we have

$$\begin{aligned} x \in f^{-1}\left(\bigcup_{\gamma\in\Gamma} B_\gamma\right) &\Leftrightarrow f(x) \in \bigcup_{\gamma\in\Gamma} B_\gamma \\ &\Leftrightarrow f(x) \in B_\gamma, \quad \text{for some} \quad \gamma \in \Gamma \\ &\Leftrightarrow x \in f^{-1}(B_\gamma), \quad \text{for some} \quad \gamma \in \Gamma \\ &\Leftrightarrow x \in \bigcup_{\gamma\in\Gamma} f^{-1}(B_\gamma) \end{aligned}$$

Thus, we have proved $f^{-1}(\bigcup_{\gamma\in\Gamma} B_\gamma) = \bigcup_{\gamma\in\Gamma} f^{-1}(B_\gamma)$.

(b) Replacing $\bigcup$ by $\bigcap$ and the phrase "for some" by "for all" in the proof of part (a) yields a proof for part (b). The student should write down each change, step by step, until he is fully convinced.

Theorem 12. Let $f: X \to Y$ be a function and let B and C be any subsets of Y. Then

$$f^{-1}(B - C) = f^{-1}(B) - f^{-1}(C)$$

Proof. Let us examine the following equivalences:

$$\begin{aligned} x \in f^{-1}(B - C) &\Leftrightarrow f(x) \in B - C && \text{Def. 9} \\ &\Leftrightarrow f(x) \in B \wedge f(x) \notin C && \text{Def. 5 (Ch. 2)} \\ &\Leftrightarrow x \in f^{-1}(B) \wedge x \notin f^{-1}(C) && \text{Def. 9} \\ &\Leftrightarrow x \in [f^{-1}(B) - f^{-1}(C)] && \text{Def. 5 (Ch. 2)} \end{aligned}$$

This proves that

$$f^{-1}(B - C) = f^{-1}(B) - f^{-1}(C)$$

EXERCISE 3.5

1. In Problem 2, Exercise 3.4, find
   (a) $f(\{-1, 0, 1\})$, $f(\{\sqrt{2}, \pi\})$, and $f(\{2, \log 2\})$
   (b) $f^{-1}(\{0, 1\})$, $f^{-1}(\{-3, 3\})$, $f^{-1}(\{4, 5\})$, and $f^{-1}(\{-3, 4, 5\})$.
2. In Problem 3, Exercise 3.4, find
   (a) $f(\{-7, 3, 6\})$, $f(\{-8, 2, 7\})$, and $f(\{-9, 1, 8\})$
   (b) $f^{-1}(\{0, 1\})$, $f^{-1}(\{-3, 3\})$, and $f^{-1}(\{1, 2, 3\})$.
3. In Problem 4, Exercise 3.4, find $f(\{v, w\})$, $f^{-1}(\{c\})$, and $f^{-1}(\{a, b\})$.
4. Let $f: X \to Y$ be a function, and let $A \subseteq X$, $B \subseteq Y$. Prove that
   (a) $A \subseteq f^{-1}(f(A))$

(b) $f(f^{-1}(B)) \subseteq B$.

5. Let $f: X \to Y$ be a function, and let $A \subseteq X$, $B \subseteq Y$. Find examples which show that the following statements are false.
   (a) if $B \neq \varnothing$, then $f^{-1}(B) \neq \varnothing$
   (b) $f^{-1}(f(A)) = A$
   (c) $f(f^{-1}(B)) = B$
   (d) $f(X) = Y$
6. Prove that Problem 5(c) is true if $f(X) = Y$.
7. Let $f: X \to Y$ be a function such that $f(X) = Y$, and let B and C be subsets of Y. Prove that $B = C$ if $f^{-1}(B) = f^{-1}(C)$. Give an example which shows that the assertion is false if $f(X) \neq Y$.
8. Let X and Y be two sets, and let $p_X: X \times Y \to X$ and $p_Y: X \times Y \to Y$ be two functions given respectively by $p_X(x, y) = x$, $p_Y(x, y) = y$ for all $(x, y) \in X \times Y$ (p_X and p_Y are called the *X-projection* and the *Y-projection*, respectively). Prove that if $\mathscr{R}$ is a relation from X to Y, that is, $\mathscr{R} \subseteq X \times Y$, then $p_X(\mathscr{R}) = \text{Dom}\,\mathscr{R}$ and $p_Y(\mathscr{R}) = \text{Im}\,\mathscr{R}$.
9. Let $f: X \to Y$ be a function, and let $A \subseteq X$, $B \subseteq Y$. Prove that
   (a) $f(A \cap f^{-1}(B)) = f(A) \cap B$
   (b) $f(f^{-1}(B)) = f(X) \cap B$.
10. Let $f: X \to Y$ be a function, and let $B \subseteq Y$. Prove that
$$f^{-1}(Y - B) = X - f^{-1}(B)$$
11. Let $f: X \to Y$ be a function, and let A and B be subsets of X. Give an example which shows that it is not true that
$$f(A - B) = f(A) - f(B)$$
12. Prove Theorem 9.
13. Let $f: X \to Y$ be a function. Prove that the family $\mathscr{P} = \{f^{-1}(y) \mid f^{-1}(y) \neq \varnothing, y \in Y\}$ of sets forms a partition of X.
14. Let $f: X \to Y$ be a function. Prove that the relation $\mathscr{R}(f) = \{(a, b) \in X \times X \mid f(a) = f(b)\}$ is an equivalence relation of X.
15. Let f, $\mathscr{P}$ and $\mathscr{R}(f)$ be as in Problems 13 and 14. Prove that $X/\mathscr{R}(f) = \mathscr{P}$ and $X/\mathscr{P} = \mathscr{R}(f)$.

6. INJECTIVE, SURJECTIVE, AND BIJECTIVE FUNCTIONS

In the study of functions, we find it convenient to give names to three important types of functions

Definition 10. A function $f: X \to Y$ is said to be *injective* or *one-to-one* provided that if $x_1, x_2 \in X$ with $f(x_1) = f(x_2)$ then $x_1 = x_2$. An injective function is called an *injection*.

By the Contrapositive Law in logic, we may say equivalently that the function $f: X \to Y$ is an injection if and only if $x_1, x_2 \in X$ with $x_1 \neq x_2$ implies that $f(x_1) \neq f(x_2)$. For example, the inclusion function in Problem 7, Exercise 3.4, is an injection.

Definition 11. A function $f: X \to Y$ is said to be *surjective* or *onto* provided that if $y \in Y$, then there exists at least one $x \in X$ such that $f(x) = y$. A surjective function is called a *surjection*. In other words, $f: X \to Y$ is a surjection if and only if $f(X) = Y$.

The function in Example 7, Section 4, for instance, is not surjective.

EXAMPLE 12. The sine function $f: \mathbf{R} \to [-1, 1]$ given by $f(x) = \sin x$ is a surjection; but if the range $[-1, 1]$ is replaced by $\mathbf{R}$, then $f: \mathbf{R} \to \mathbf{R}$ is not surjective.

Definition 12. A function $f: X \to Y$ is called a *bijection* or said to be *bijective* if it is both injective and surjective. A bijection is also called a *one-to-one correspondence.*

For example, the identity function in Example 9, Section 4, is a bijection. Definition 10, 11, and 12 are illustrated in three diagrams on the following page. Sets X and Y are represented as sets of dots within the circles. In each picture, every dot in X is paired with some dot in Y by an arrow drawn between them. The set of pairs so obtained gives rise to a function $f: X \to Y$.

For an injection, the result of Theorem 10(b) may be improved.

Theorem 13. Let $f: X \to Y$ be an injection and let $\{A_\gamma \mid \gamma \in \Gamma\}$ be a family of subsets of X. Then

$$f\left(\bigcap_{\gamma \in \Gamma} A_\gamma\right) = \bigcap_{\gamma \in \Gamma} f(A_\gamma)$$

Proof. By Definition 9, and Definition 7 of Chapter 2, we have

$$y \in \bigcap_{\gamma \in \Gamma} f(A_\gamma) \Leftrightarrow y \in f(A_\gamma) \forall \gamma \in \Gamma$$

$$\Leftrightarrow (\exists x_\gamma \in A_\gamma \text{ such that } y = f(x_\gamma)) \forall \gamma \in \Gamma$$

Since $f: X \to Y$ is injective, all these x_γ's are the same; we denote this element by x_0. Then we have

$$\begin{aligned} y \in \bigcap_{\gamma\in\Gamma} f(A_\gamma) &\Leftrightarrow \exists x_0 \in A_\gamma \quad \text{such that} \quad y = f(x_0), \forall \gamma \in \Gamma \\ &\Leftrightarrow \exists x_0 \in \bigcap_{\gamma\in\Gamma} A_\gamma \quad \text{such that} \quad y = f(x_0) \\ &\Leftrightarrow y \in f\left(\bigcap_{\gamma\in\Gamma} A_\gamma\right) \end{aligned}$$

Therefore, $f(\bigcap_{\gamma\in\Gamma} A_\gamma) = \bigcap_{\gamma\in\Gamma} f(A_\gamma)$.

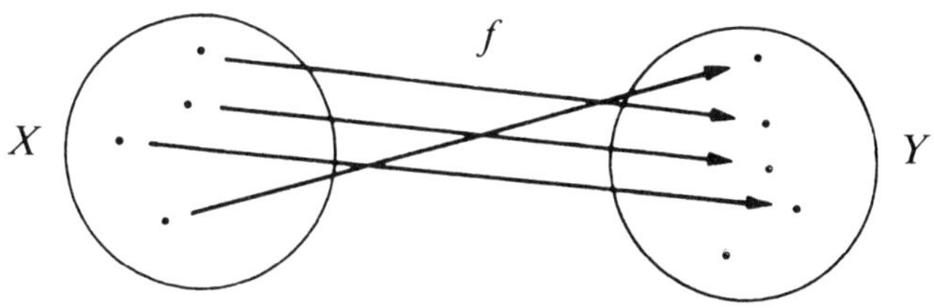

Figure 12. $f: X \to Y$ is injective

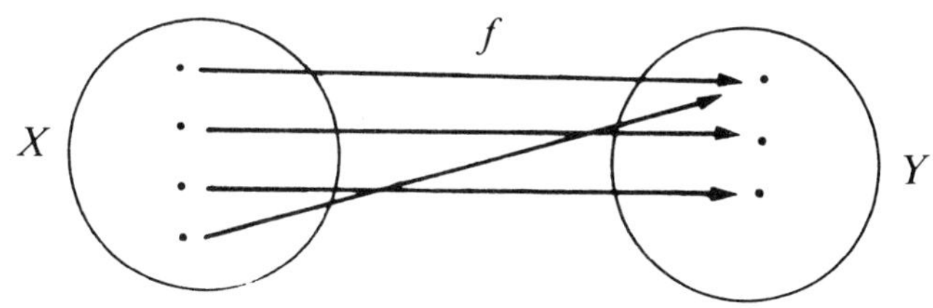

Figure 13. $f: X \to Y$ is surjective

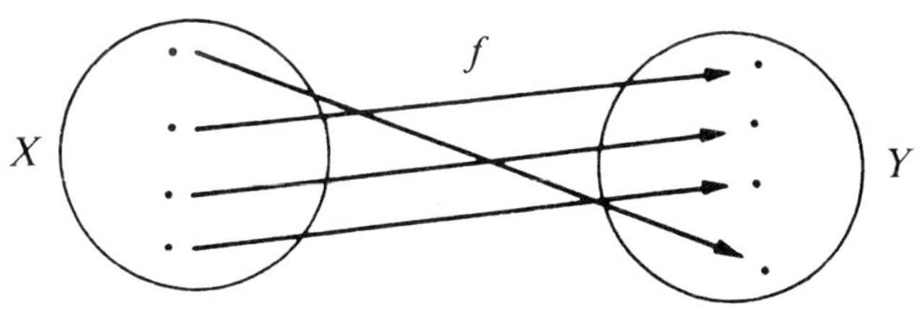

Figure 14. $f: X \to Y$ is bijective

Recall that if $\mathcal{R}$ is a relation from X to Y, then its inverse

$$\mathcal{R}^{-1} = \{(y, x) \mid (x, y) \in \mathcal{R}\}$$

is a relation from Y to X. Since a function $f: X \to Y$ is a (particular kind of) relation from X to Y, f^{-1} is at least a relation from Y to X. It is natural to inquire when f^{-1} becomes a function. This question is considered in the following theorem.

Theorem 14. Let $f: X \to Y$ be a bijection. Then $f^{-1} : Y \to X$ is a bijection.

Proof. We shall first prove that the relation f^{-1} from Y to X forms a function. Since $f: X \to Y$ is surjective, by Problem 3(a), Exercise 3.2, we have $\text{Dom}(f^{-1}) = \text{Im}(f) = Y$. Thus, condition (a) of Definition 8 is satisfied. To show that f^{-1} satisfies the other condition, we let $(y, x_1) \in f^{-1}$ and $(y, x_2) \in f^{-1}$. Then we have $(x_1, y) \in f$ and $(x_2, y) \in f$. Consequently $f(x_1) = f(x_2)$. Now, because $f: X \to Y$ is injective, the last equality implies $x_1 = x_2$. Thus we have established that $f^{-1} : Y \to X$ is a function.

To show that the function $f^{-1} : Y \to X$ is injective, let $y_1, y_2 \in Y$ with $f^{-1}(y_1) = f^{-1}(y_2) = x$ (say). Then we have $f(x) = y_1$ and $f(x) = y_2$, and hence $y_1 = y_2$. This proves that f^{-1} is injective.

Finally, it remains to be shown that $f^{-1} : Y \to X$ is surjective. By Problem 3(b) of Exercise 3.2, we have $\text{Im}(f^{-1}) = \text{Dom}(f) = X$, which proves that f^{-1} is surjective. Thus the proof is complete.

If $f: X \to Y$ is a bijection, the function $f^{-1} : Y \to X$ is called the *inverse function* of f (see also Problem 17, Exercise 3.6).

By virtue of Theorem 14, if $f: X \to Y$ is a bijection (=one-to-one correspondence), we shall say that f is a one-to-one correspondence *between* the sets X and Y.

EXERCISE 3.6

1. Which of the functions in Problem 2, 3, and 4 of Exercise 3.4 are injective? Surjective?
2. Which of the functions in Problem 5 and 6 of Exercise 3.4 are injective? Surjective? Bijective?
3. Let $f: \mathbf{R} \to \mathbf{R}$ be the function defined by $f(x) = 3x - 2$, for all $x \in \mathbf{R}$.
   (a) Prove that the function f is a bijection.
   (b) Find the inverse f^{-1} of f.
4. Let $g: (-\pi/2, \pi/2) \to \mathbf{R}$ be the function given by $g(x) = \tan x$, for all $-\pi/2 < x < \pi/2$. Is this function bijective? If so, describe its inverse function.

5. Prove that the characteristic function $\chi_A : X \to \{0, 1\}$ in Example 8, Section 4, is surjective if and only if $A \subset X$ and $A \neq \varnothing$. When does $\chi_A : X \to \{0, 1\}$ become an injection?
6. Prove that the constant function $C_b : X \to Y$ is surjective if and only if $Y = \{b\}$. When does $C_b : X \to Y$ become an injection?
7. Prove that the X-projection $p_X : X \times Y \to X$ and the Y-projection $p_Y : X \times Y \to Y$ in Problem 8, Exercise 3.5, are surjective. When is the X-projection an injection?
8. Prove that there is a one-to-one correspondence between the set **N** of natural numbers and the set of all even natural numbers.
9. Prove that there is a one-to-one correspondence between the set **Z** of integers and the set of all odd integers.
10. Let X be a finite set with m elements and let Y be a finite set with n elements. Prove that
    (a) If $m > n$, then there can be no injection $f: X \to Y$.
    (b) If $m \leq n$, there exist exactly $n!/(n-m)!$ injections.
    [See also Problem 11, Exercise 3.4]
11. Let X be a finite set with m elements. How many bijections from X onto X exist? [Remark: A bijection from a finite set onto itself is sometimes called a *permutation*.]
12. Let $f: X \to Y$ be a function, and let $A \subseteq X$, $B \subseteq Y$. Prove that
    (a) If f is injective, then $f^{-1}(f(A)) = A$.
    (b) If f is surjective, the $f(f^{-1}(B)) = B$.
13. Prove the following converse of Problem 12:
    (a) If $f^{-1}(f(A)) = A$ for all $A \subset X$, then f is injective.
    (b) If $f(f^{-1}(B)) = B$ for all $B \subset Y$, then f is surjective.
14. Let $f: X \to Y$ be an injection and let $B \subset X$. Prove that $f(X - B) = f(X) - f(B)$.
15. Let $f: X \to Y$ be an injection, and let A and B be subsets of X. Prove that $f(A - B) = f(A) - f(B)$. [Compare this with Problem 11 of Exercise 3.5]
16. Let $f: X \to X$ be a function such that $f(f(x)) = x$ for all $x \in X$. Prove that f is a symmetric relation on X.
17. Prove the following converse of Theorem 14: Let $f: X \to Y$ be a function such that f^{-1} is a function from Y to X. Then $f: X \to Y$ is bijective.
18. Prove that $f: X \to Y$ is injective if and only if $f(A \cap B) = f(A) \cap f(B)$ for all subsets A and B of X.
19. Prove the following converse of Problem 15:
    If $f(A - B) = f(A) - f(B)$ for all subsets A and B of X, then f is injective.
20. Is the converse of Problem 14 true?

7. COMPOSITION OF FUNCTIONS

A function $f:X \to Y$ may be considered as a machine that takes an arbitrary object x of the set X, operates on it in a certain way, and transforms if into a new object $f(x)$, an output of the machine. This idea is illustrated in Figure 15.

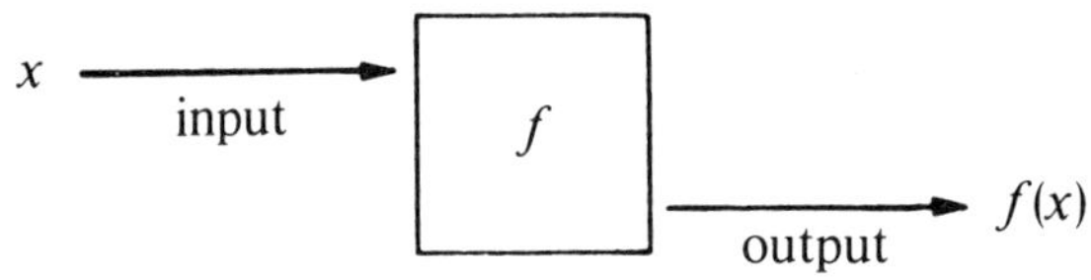

Figure 15.

Let $f: X \to Y$ and $g: Y \to Z$ be two given functions, where the domain of the second function is the same as the range of the first function. Imagine these two functions as two machines such as a washer and a dryer. We don't have to be inventors to imagine the possibility of combining these two machines into one new machine; the result would be a washer-dryer combination that takes a dirty garment x, washes it so that it becomes a clean but wet garment $f(x)$, and then dries it. The outcome is a clean and dry garment $g(f(x))$. The idea is illustrated in Figure 16.

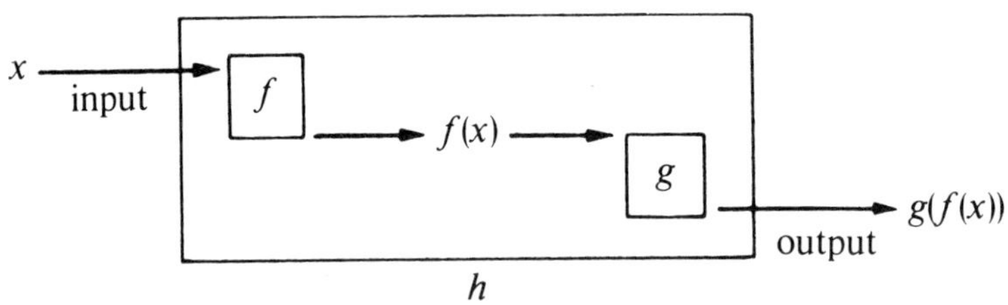

Figure 16.

The "combination" of the machines $f: X \to Y$ and $g: Y \to Z$ results in a new machine, denoted by $h: X \to Z$, which takes an arbitrary object x in X and transforms it into the object $h(x) = g(f(x))$ in Z. The traditional notation for h is $g \circ f$ and $(g \circ f)(x) = g(f(x))$; the traditional name for the term "combination" is "composition."

Now we are ready for the following definition.

Definition 13. Let $f: X \to Y$ and $g: Y \to Z$ be two functions. The *composition* of these two functions is the function $g \circ f: X \to Z$ where $(g \circ f)(x) = g(f(x))$ for all x in X. In another notation

$$g \circ f = \{(x, z) \in X \times Z \mid \exists y \in Y \text{ such that } (x, y) \in f \wedge (y, z) \in g\}$$

EXAMPLE 13. Let $f: \mathbf{R} \to \mathbf{R}$ and $g: \mathbf{R} \to \mathbf{R}$ be two functions given respectively by $f(x) = x + 1$ and $g(x) = x^2$ for all x in $\mathbf{R}$. Find the composition $(g \circ f)(x)$ and $(f \circ g)(x)$.

Solution. Using Definition 13, we have

(1)
$$\begin{aligned}(g \circ f)(x) &= g(f(x)) \\ &= g(x + 1) \\ &= (x + 1)^2 \\ &= x^2 + 2x + 1\end{aligned}$$

(2)
$$\begin{aligned}(f \circ g)(x) &= f(g(x)) \\ &= f(x^2) \\ &= x^2 + 1\end{aligned}$$

The result of Example 13 tells us that in general $g \circ f \neq f \circ g$; therefore, *functional composition is not commutative.*

Theorem 15. Functional composition is associative. That is, if $f: X \to Y$. $g: Y \to Z$, and $h: Z \to W$, then

$$(h \circ g) \circ f = h \circ (g \circ f)$$

Proof. We first note that both $h \circ (g \circ f)$ and $(h \circ g) \circ f$ give rise to functions from X to W. Therefore, to show that $h \circ (g \circ f) = (h \circ g) \circ f$, by Theorem 7 of Section 4, we need only to show that $[h \circ (g \circ f)](x) = [(h \circ g) \circ f](x)$ for all x in X. We use Definition 13 to derive the following:

$$[h \circ (g \circ f)](x) = h(g \circ f(x)) = h(g(f(x)))$$

and

$$[(h \circ g) \circ f](x) = (h \circ g)(f(x)) = h(g(f(x)))$$

for all x in X. This shows that $[h \circ (g \circ f)](x) = [(h \circ g) \circ f](x), \forall x \in X$. The proof is now complete.

Theorem 16. Let $f: X \to Y$ be a function. Then

(a) If there exists a function $g: Y \to X$ such that $g \circ f = I_X$ (where $I_X: X \to X$ is the identity function defined in Example 9, Section 4), then $f: X \to Y$ is injective.

(b) If there exists a function $h: Y \to X$ such that $f \circ h = I_Y$, then $f: X \to Y$ is surjective.

Proof. (a) Suppose that there exists a function $g: Y \to X$ such that $g \circ f = I_X$. Then for any x_1 and x_2 in X with $f(x_1) = f(x_2)$, we have

$$x_1 = (g \circ f)(x_1) = g(f(x_1)) = g(f(x_2)) = (g \circ f)(x_2) = x_2$$

This proves that $f: X \to Y$ is injective.

(b) Suppose that there exists a function $h: Y \to X$ such that $f \circ h = I_Y$. Then for each $y \in Y$, there exists an element

$$x = h(y) \in X$$

such that

$$f(x) = f(h(y)) = (f \circ h)(y) = I_Y(y) = y$$

By Definition 11, $f: X \to Y$ is surjective.

EXERCISE 3.7

1. Let $f: \mathbf{R} \to \mathbf{R}$ and $g: \mathbf{R} \to \mathbf{R}$ be two functions defined by $f(x) = 2x^3 + 1$ and $g(x) = \cot x$, respectively, for all $x \in \mathbf{R}$.
   (a) Find the composition $g \circ f$.
   (b) Find the composition $f \circ g$.
2. Let $f: \mathbf{R}_+ \to \mathbf{R}$ and $g: \mathbf{R} \to \mathbf{R}_+$ be two functions defined by $f(x) = \log_{10} x$ and $g(x) = 10^x$, respectively, for all $x \in \mathbf{R}$.
   (a) Find the composition $g \circ f: \mathbf{R}_+ \to \mathbf{R}_+$.
   (b) Find the composition $f \circ g: \mathbf{R} \to \mathbf{R}$.
3. Let f, g, and h be the functions given in Problem 6, Exercise 3.4.
   (a) Find the composition $g \circ f$.
   (b) Find the composition $h \circ g$.
   (c) Find the composition $h \circ (g \circ f)$.
   (d) Find the composition $(h \circ g) \circ f$.
   (e) Compare your answers for $h \circ (g \circ f)$ and $(h \circ g) \circ f$; are they the same?
4. Let $f: X \to Y$ be a function. Prove that $f \circ I_X = f = I_Y \circ f$.
5. Let $f: X \to Y$ be a bijection and let $f^{-1}: Y \to X$ be the inverse function of f. Prove that $f^{-1} \circ f = I_X$ and $f \circ f^{-1} = I_Y$.
6. Let $f: X \to Y$ be a function. If there exist functions $g: Y \to X$ and

$h: Y \to X$ such that $g \circ f = I_X$ and $f \circ h = I_Y$, prove that $f: X \to Y$ is bijective and that $g = h = f^{-1}$.

7. Let $f: X \to Y$ and $g: Y \to Z$ be functions. Prove that
   (a) If $f: X \to Y$ and $g: Y \to Z$ are injective, then so is $g \circ f: X \to Z$.
   (b) If $f: X \to Y$ and $g: Y \to Z$ are surjective, then so is $g \circ f: X \to Z$.
8. Let $\mathscr{R}$ be a relation from X to Y and let $\mathscr{S}$ be a relation from Y to Z. We may, as in the composition of functions, define the *composition* of these relations by

$$\mathscr{S} \circ \mathscr{R} = \{(x, z) \in X \times Z \mid (\exists y)[(x, y) \in \mathscr{R} \wedge (y, z) \in \mathscr{S}]\}$$

   which is a relation from X to Z. Prove that
   (a) $(\mathscr{S} \circ \mathscr{R})^{-1} = \mathscr{R}^{-1} \circ \mathscr{S}^{-1}$.
   (b) If furthermore $\mathscr{T}$ is a relation from Z to W, then $\mathscr{T} \circ (\mathscr{S} \circ \mathscr{R}) = (\mathscr{T} \circ \mathscr{S}) \circ \mathscr{R}$.
9. Let $\mathscr{R}$ be a relation on X. Prove that
   (a) If $\mathscr{R} \circ \mathscr{R} \supseteq I_X$, then $\mathscr{R}$ is symmetric.
   (b) $\mathscr{R}$ is transitive if and only if $\mathscr{R} \circ \mathscr{R} \subseteq \mathscr{R}$.
10. Let $\mathscr{R}$ be a relation from X to Y and let $\mathscr{S}$, $\mathscr{T}$ be relations from Y to Z. Prove that

$$(\mathscr{T} \cup \mathscr{S}) \circ \mathscr{R} = (\mathscr{T} \circ \mathscr{R}) \cup (\mathscr{S} \circ \mathscr{R}).$$

11. Let $\mathscr{R}$, $\mathscr{S}$ be relations from X to Y and let $\mathscr{T}$ be a relation from Y to Z. Prove that

$$\mathscr{T} \circ (\mathscr{R} \cap \mathscr{S}) \subseteq (\mathscr{T} \circ \mathscr{R}) \cap (\mathscr{T} \circ \mathscr{S}).$$

12. Let $f: X \to Y$ and $g: Y \to Z$ be two bijections. Prove that $g \circ f: X \to Z$ is a bijection, and that the inverse function $(g \circ f)^{-1}: Z \to X$ is the same as the composition $f^{-1} \circ g^{-1}: Z \to X$ of the inverse functions $g^{-1}: Z \to Y$ and $f^{-1}: Y \to X$. That is, $(g \circ f)^{-1} = f^{-1} \circ g^{-1}$.

Chapter Four

Boolean Algebra with Applications

An interesting and useful observation of sets together with the operations of union, intersection and complementation lead to an abstraction of an important and useful algebraic system known as Boolean algebra, named in honor of the logician and mathematician George Boole (1813-1864). In this chapter, the basic properties of a Boolean algebra are developed. Applications to electronic circuits and to digital circuit design are given.

1. BOOLEAN ALGEBRA

There are many seemingly different but equivalent definitions of a Boolean algebra. Because of its similarity to the structure of a certain set of sets, we shall adopt the following definition. By a *binary operation* on a set S we mean a function $\circ$ from the Cartesian square $S \times S$ to S. For any $(a, b) \in S \times S$, it is customary to write $a \circ b$ for $\circ(a, b)$.

Definition 1. A *Boolean algebra* is a triplet $(\mathscr{B}, +, *)$ in which $\mathscr{B}$ is a nonempty set, $+$ and $*$ are two binary operations on $\mathscr{B}$ satisfying postulates (a) through (e), where $a, b, c \in \mathscr{B}$:

(a) Commutativities: $a + b = b + a$
$a * b = b * a$

(b) Associativities: $a + (b + c) = (a + b) + c$
$a * (b * c) = (a * b) * c$

(c) Distributivities: $a * (b + c) = (a * b) + (a * c)$
$a + (b * c) = (a + b) * (a + c)$

(d) Unities: There exist $0, 1 \in \mathscr{B}$ such that $0 \neq 1$,

$$a + 0 = a \quad \text{and} \quad a * 1 = a.$$

(e) Complementation: For all $a \in \mathscr{B}$, there exists $a' \in \mathscr{B}$ such that

$$a + a' = 1 \text{ and } a * a' = 0.$$

The operations $+$ and $*$ are called *sum* (OR) and *product* (AND), respectively. The elements 0 and 1 in (d), though not related in any way to the integers 0 and 1, are called *zero* and *identity* elements. The element a' in (e) is called the *complement* of a.

EXAMPLE 1. Let $\mathscr{B}_2 = \{0, 1\}$ and let the binary operations $+$ and $*$ on $\mathscr{B}_2$ be given by the following tables:

Table 1.

+	0	1
0	0	1
1	1	1

*	0	1
0	0	0
1	0	1

Then, the triplet $(\mathscr{B}_2, + \ *)$ satisfies the definition of a Boolean algebra. Obviously, by (d), this is a Boolean algebra with the smallest number of elements. What are the elements $0'$ and $1'$?

EXAMPLE 2. In the power set $\mathscr{P}(U)$ of an arbitrarily given nonempty set U, let $A + B = A \cup B$ and $A * B = A \cap B$, for all $A, B \in \mathscr{P}(U)$. Then $(\mathscr{P}(U), +, *)$ is a Boolean algebra.

EXAMPLE 3. Let $\mathscr{B}$ be the set of logical equivalence classes of logical statements generated by a set $\{p, q, r, \ldots\}$ of simple statements. Then the triplet $(\mathscr{B}, \vee, \wedge)$ is a Boolean algebra in which 0 is the class of contradictions and 1 is the class of tautologies. What is a' for a given $a \in \mathscr{B}$?

We are now ready to prove some basic theorems on a Boolean algebra.

Theorem 1. In a Boolean algebra,

(a) 0 and 1 are unique.

(b) For each a in $\mathscr{B}$, a' is unique.

Proof. (a) Let $\tilde{0}$ be possibly another zero element in $\mathscr{B}$. Then

$\tilde{0} = \tilde{0} + 0$	Def. 1(d)
$= 0 + \tilde{0}$	Def. 1(a)
$= 0$	Def. 1(d), since $\tilde{0}$ is a zero element.

The proof for the uniqueness of 1 is similar.

(b) Let x be any element in $\mathscr{B}$ such that $a + x = 1$ and $a * x = 0$. Then

$x = x * 1$	Def. 1(d)
$= x * (a + a')$	Def. 1(e)
$= (x * a) + (x * a')$	Def. 1(c)
$= (a * x) + (x * a')$	Def. 1(a)
$= 0 + (x * a')$	Since $a * x = 0$
$= x * a'$	Def. 1(c) and 1(d)

Similary, it can be proved that $a' = x * a'$. Thus, $x = a'$ and hence a' is unique.

Theorem 2. $(a')' = a$ for all a in $\mathscr{B}$.

The proof of this theorem is left as an exercise (See Exercise 3).

Theorem 3. The operations $+$ and $*$ are idempotent:
$a + a = a$ and $a * a = a$ for all a in $\mathscr{B}$.

Proof.

$a + a = (a + a) * 1$	Def. 1(d)
$= (a + a) * (a + a')$	Def. 1(e)
$= a + (a * a')$	Def. 1(c)
$= a + 0$	Def. 1(e)
$= a$	Def. 1(d)

The proof for $a * a = a$ is left to the reader.

Theorem 4. In a Boolean algebra, $0' = 1$ and $1' = 0$.

Proof. Since $0 + 1 = 1 + 0 = 1$, by (a) and (d) of Def. 1, and $0 * 1 = 0$, by Def. 1(d), we have $0' = 1$ by Def. 1(e) and Theorem 1(b). The second equation $1' = 0$ can be proved similarly, or by applying Theorem 2 to $0' = 1$.

Theorem 5. $a + 1 = 1$ and $a * 0 = 0$ for all a in $\mathscr{B}$.

Proof.

$a + 1 = a + (a + a')$	Def. 1(e)
$= (a + a) + a'$	Def. 1(b)

$= a + a'$	Theorem 3
$= 1$	Def. 1(e)

The proof for $a * 0 = 0$ is left as an exercise.

Theorem 6. De Morgan's Laws: $(a + b)' = a' * b'$ and $(a * b)' = a' + b'$ for all a and b in $\mathscr{B}$.

Proof. By Def. 1(e) and Theorem 1(b), to prove the first half, it is sufficient to show $(a + b) + (a' * b') = 1$ and $(a + b) * (a' * b') = 0$.
We have,

$(a + b) + (a' * b') = [(a + b) + a'] * [(a + b) + b']$	Def. 1(c)
$= [a' + (a + b)] * [a + (b + b')]$	Def. 1(a) and 1(b)
$= [(a' + a) + b] * [a + 1]$	Def. 1(b) and 1(e)
$= (1 + b) * (a + 1)$	Def. 1(e)
$= 1 * 1$	Def. 1(a), Th. 5
$= 1$	

The proofs for $(a + b) * (a' * b') = 0$ and for the second half of the theorem are left as exercises.

The Boolean algebra $(\mathscr{P}(U), \cup, \cap)$ plays a very important role in the study of abstract Boolean algebras. If $(\mathscr{B}, +, *)$ is a finite Boolean algebra, then it can be proved that there exists a finite set U such that $(\mathscr{P}(U), \cup, \cap)$ has exactly the same algebraic properties as those of $(\mathscr{B}, +, *)$ so that $(\mathscr{P}(U), \cup, \cap)$ can be a perfect substitute for $(\mathscr{B}, +, *)$. This is not exactly true if $\mathscr{B}$ is infinite, but $(\mathscr{B}, +, *)$ can be identified with a subfamily of $\mathscr{P}(U)$ for a suitable infinite set U.

EXERCISE 4.1

1. Verify that $(\mathscr{B}_2, +, *)$ in Example 1 is the smallest Boolean algebra.
2. Verify that $(\mathscr{P}(U), \cup, \cap)$ in Example 2 satisfies postulates (a) through (e) of Definition 1.
3. Prove Theorem 2: $(a')' = a$ for all $a \in \mathscr{B}$.
4. Can a Boolean algebra contain exactly three elements? Why?
5. Can any Boolean algebra contain any element a such that $a' = a$?
6. Prove that $a * a = a$ for all a in $\mathscr{B}$.
7. Prove that $1' = 0$.
8. Prove that $a * 0 = 0$.
9. Prove the following *absorption laws*: If $a, b \in \mathscr{B}$ then
   (a) $a + (a * b) = a$
   (b) $a * (a + b) = a$.

10. Prove that $(a+b)*(a'*b')=0$.
11. Prove the second part of Theorem 6: $(a*b)'=a'+b'$.
12. Prove that if $a+b=0$ then $a=0$ and $b=0$.
13. Prove that if $a*b=1$ then $a=1$ and $b=1$.
14. Prove that $a*(a'+b)=a*b$.
15. Let a and b be any elements in a Boolean algebra. Prove that the following four conditions are equivalent:
    (a) $a+b=b$
    (b) $a'+b=1$
    (c) $a*b=a$
    (d) $a*b'=0$.
16. In a Boolean algebra $\mathscr{B}$, a is said to *precede* b, denoted $a \le b$, if and only if one of the four conditions in Problem 15, above, is satisfied. Prove that:
    (a) $\le$ is reflexive: $a \le a$ for all a in $\mathscr{B}$
    (b) $\le$ is transitive: if $a \le b$ and $b \le c$ then $a \le c$
    (c) $\le$ is antisymmetric: if $a \le b$ and $b \le a$ then $a=b$.
    [Such a relation is called a *partial order*.]
17. Prove that $0 \le a \le 1$ for all a in $\mathscr{B}$.

2. BOOLEAN FUNCTIONS

A variable which takes the values of 0 or 1 is called a *binary variable*. A *Boolean function* is a formal expression formed with binary variables, the two binary operations $+$, $*$, the unary operation $'$ and parentheses. We shall always use the operation tables of the Boolean algebra $\mathscr{B}_2$ of Example 1 in evaluating the value of a Boolean function. Thus, whenever values of the variables are given, the Boolean function must take values 0 or 1. The logical symbols p, q, r, . . . used in Chapter 1 are binary variables with values 0 (for false) and 1 (for true). Any logical expression, such as $p \vee (q \wedge r)'$, is a Boolean function, where $\vee$, $\wedge$ and $'$ are respectively the sum, product, and complement in the Boolean algebra of Example 3.

A *truth table* for a Boolean function is a table resembling a truth table of a logical statement, except the values 0 and 1 are used instead of F and T. There is a minor difference between the truth tables of Boolean functions and the truth tables of logic: in logic, truth tables go from T to F, whereas in Boolean functions, the truth tables go from 0 to 1. The main reason for adopting "backward truth tables" is to conform with the order in the binary coding of decimal numbers.

For simplicity, we shall henceforth use juxtaposition to record the

Boolean product $*$ (or $\wedge$). Parenthesis will be omitted whenever the Associative Law applies. Thus, we shall write, for example, xyz instead of $x*(y*z)$ or $(x \wedge y) \wedge z$.

EXAMPLE 4. The truth tables of the Boolean functions $F = x + y'z$, $G = x'yz$ and $H = xy' + x'z$ are combined in one large table with the last three columns representing the values of F, G, and H as indicated:

Table 2

x	y	z	F	G	H
0	0	0	0	0	0
0	0	1	1	0	1
0	1	0	0	0	0
0	1	1	0	1	1
1	0	0	1	0	1
1	0	1	1	0	1
1	1	0	1	0	0
1	1	1	1	0	0

Boolean functions can often be simplified by minimizing the number of "literals" and the number of terms by using $x + x' = 1$, $xx' = 0$ and other identities. A *literal* is a primed (complemented) or unprimed variable. In applications of the theory of Boolean algebra to circuit designs (see Sections 3 and 4) the minimization of the number of literals means that a saving of materials or equipment can be achieved.

EXAMPLE 5. Simplify the following Boolean functions:

(a) $xy + xy' + x'y'$

(b) $xy + xz + y'z$

(c) $x + x'y$.

Solution. (a)

$$\begin{aligned} xy + xy' + x'y' &= x(y + y') + x'y' && \text{Def. 1(c)} \\ &= x + x'y' && y + y' = 1 \\ &= (x + x')(x + y') && \text{Def. 1(c)} \\ &= x + y' && x + x' = 1 \end{aligned}$$

(b)

$$\begin{aligned} xy + xz + y'z &= xy + (y + y')xz + y'z && y + y' = 1 \\ &= xy + xyz + xy'z + y'z && \text{Def. 1(a), and 1(c)} \\ &= xy(1 + z) + (x + 1)y'z && \text{Def. 1(c)} \\ &= xy + y'z && \text{Th. 5} \end{aligned}$$

Simplification of (c) is left to the reader as an exercise.

In practical applications, such as designing digital circuits, one usually constructs a Boolean function from a given truth table. Such a problem has a systematic solution. For illustration, let us limit ourselves to three variables x, y, and z so that the truth tables require only 8 $(=2^3)$ rows:

Table 3

	x	y	z	$F = F(x, y, z)$
Row 1:	0	0	0	•
Row 2:	0	0	1	•
Row 3:	0	1	0	•
Row 4:	0	1	1	•
Row 5:	1	0	0	•
Row 6:	1	0	1	•
Row 7:	1	1	0	•
Row 8:	1	1	1	•

Case 1. The given Boolean function has the value 1 in row j, $1 \leq j \leq 8$, and 0 in all other rows.

To solve this simplified problem, we propose the following *basic* Boolean functions:

$$\begin{aligned} G_1 &: x'y'z' \\ G_2 &: x'y'z \\ G_3 &: x'yz' \\ G_4 &: x'yz \\ G_5 &: xy'z' \\ G_6 &: xy'z \\ G_7 &: xyz' \\ G_8 &: xyz \end{aligned}$$

It can be easily verified that each G_j, $1 \leq j \leq 8$, satisfies the requirement of having 1 in the jth row, and 0 elsewhere.

Case 2. The given function has the value 1 in rows $j_1, j_2, \ldots, j_k$, $1 \leq j_1, \ldots, j_k \leq 8$, and 0 elsewhere.

In this case, $F = G_{j_1} + G_{j_2} + \ldots + G_{j_k}$ satisfies the requirement. This can be seen easily by verifying, for example, that $G_2 + G_3 = x'y'z + x'yz'$ has the value 1 in rows 2 and 3, and 0 elsewhere.

The above process of finding a Boolean function for a given truth table is not limited to three variables. In fact the process can be used similarly for any finite number of variables. The general solution obtained in Case 2, above, is in the "sum of products" form. A different way of solving these types of problems by expressing each solution function in the "product of sums" form is indicated below in exercises 18, 19, and 20.

EXERCISE 4.2

In Problems 1 through 10, simplify each of the given Boolean functions.

1. $x + x'y$
2. $xy + x'y + x'y'$
3. $x'y + yz + xz$
4. $x(x' + y)$
5. $xy'z' + xy'z + x'z'$
6. $(x + y)(x + z)(y' + z)$
7. $x'yz + x'yz' + xy'z + xy'z'$
8. $xyz' + xy'z' + x'yz' + x'y'z'$
9. $xy + xy'z$
10. $x(y'z + yz) + yz$.

In Problems 11 through 13, simplify the given Boolean functions to the indicated number of literals.

11. $xy + zw + x + (x + z' + w')'$ to 3 literals
12. $xy + yz + xz' + yzw$ to 4 literals
13. $xyz + xyz' + x'yz + x'y'z + x'y'z'$ to 5 literals
14. Show, by truth table, that $G_2 + G_5 = x'y'z + xy'z'$ has the value 1 in rows 2 and 5, and 0 elsewhere.
15. Show that $G_2 + G_5 + G_7 = x'y'z + xy'z' + xyz'$ has the value 1 in rows 2, 5, and 7, and 0 elsewhere.
16. Give a Boolean function that has the value 0 in every row.
17. Give a Boolean function that has the value 1 in every row.
18. By taking complements of $G_1 = x'y'z'$ through $G_8 = xyz$ we have $G_1' = x + y + z, \ldots, G_8' = x' + y' + z'$. Determine the values of $G_j', j = 1, 2, \ldots, 8$.
19. Find the values of $G_1' G_3' G_4' G_6' G_8'$ for rows 1 to 8.
20. Compare the values of the Boolean functions in Problem 15 and 19, above.

3. BOOLEAN FUNCTIONS AND LOGIC GATES

With the proper input, electronic digital circuits (or logic circuits) establish logical manipulation paths. By passing binary signals through various combinations of logic circuits, any desired information for computing or control can be operated upon—where each signal represents (and is represented by) a binary variable carrying one "bit"[1] of information. Logic circuits (or *gates*) that perform the logical operations of AND (·), OR (+), and NOT (′) are shown with their symbols in Figure 1. These gates are blocks of hardware that produce a logic - 1 or logic - 0 output signal if input logic requirements are satisfied. The NOT gate is sometimes called an *inverter* since it inverts a binary signal from 0 to 1 (and from 1 to 0).

(a) Two and three input AND gates

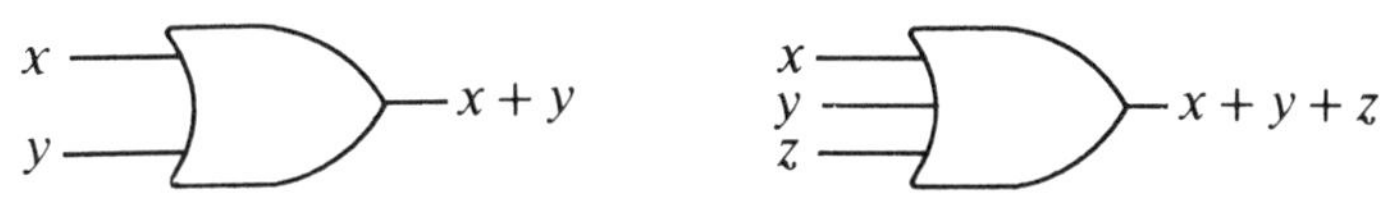

(b) Two and three input OR gates

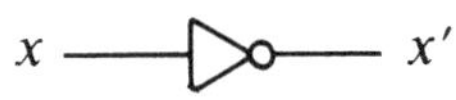

(c) An inverter (NOT gate)

Figure 1.

---

[1] A *bit*, in computer language, is a binary digit.

Since Boolean functions are formed in terms of + (OR), · (AND), and (NOT) operations, any Boolean function can be represented by an electronic circuit or a logic circuit diagram (see Example 6) using the logic gates introduced in Figure 1. We shall call the conversion of a Boolean function to a logic circuit the *implementation*.

EXAMPLE 6. The Boolean function $F = (yw + yz + xzw')'$ may be implemented in the following logic circuit diagram:

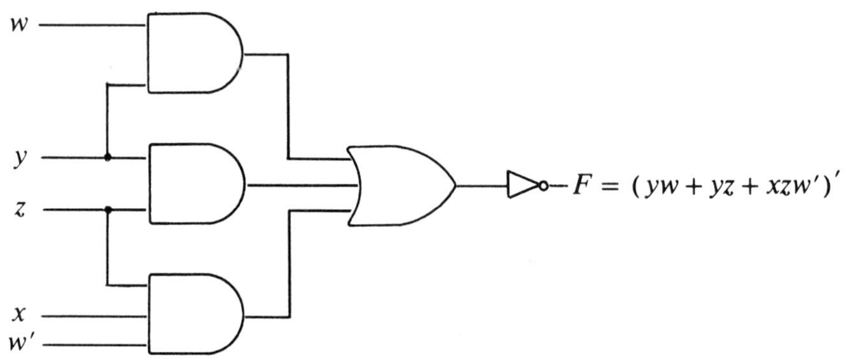

Figure 2

EXAMPLE 7. By De Morgan's Law, the Boolean function F in the previous example may be expressed as

$$F = (y' + w')(y' + z')(x' + z' + w).$$

Thus, F can also be implemented as:

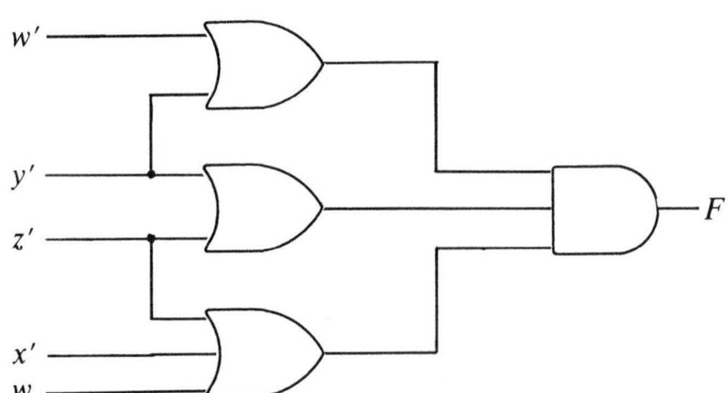

Figure 3.

Since the Boolean functions in Examples 6 and 7 have the identical truth table, the logic circuits of Figures 2 and 3 serve the identical purpose. But from the economy and efficiency points of view the circuit in Figure 3 is preferable to the one in Figure 2, because it uses one less logic gate.

One way of cutting the cost of equipment and increasing the efficiency is to create more types of useful logic gates. The following logic gates NAND, NOR, XOR (exclusive OR), and EQ (Equivalence) are created for this purpose:

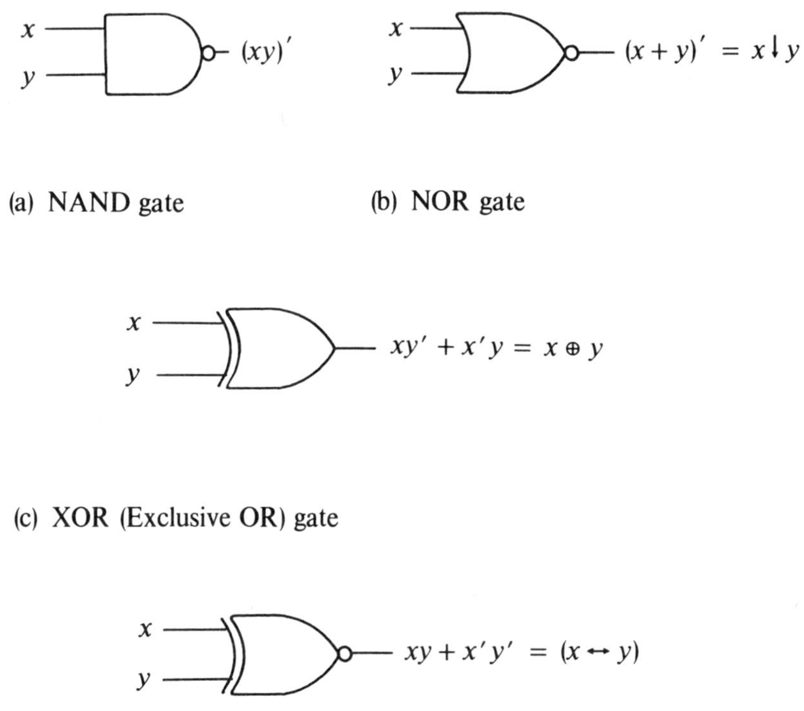

(a) NAND gate

(b) NOR gate

(c) XOR (Exclusive OR) gate

(d) EQ (Equivalence) gate

Figure 4.

Although all the four new operations (= gates) are commutative, it should be pointed out that NAND and NOR are not associative, while XOR and EQ are associative.

EXAMPLE 8. Implementations of $(x \downarrow y) \downarrow z$ and $x \downarrow (y \downarrow z)$ by NOR gates are given in Figure 5. show that $(x \downarrow y) \downarrow z$ and $x \downarrow (y \downarrow z)$ are not identical.

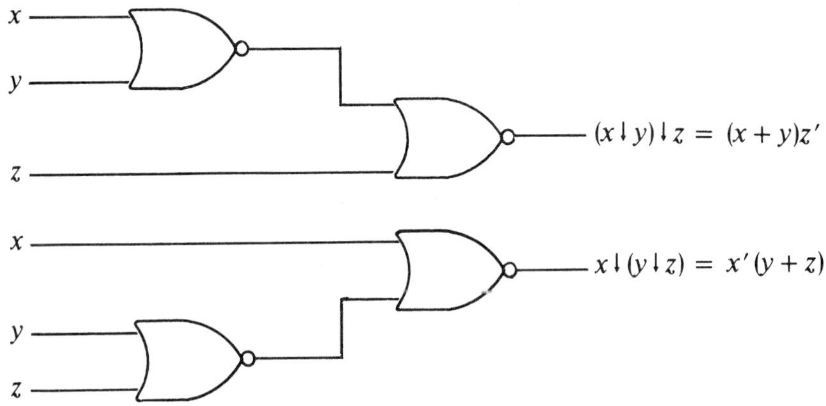

Figure 5.

Solution. For an identical set of inputs, say $x = y = 0$ and $z = 1$, the output of the upper circuit is 0, while the output of the lower circuit is 1. Therefore, the NOR gate is nonassociative.

It is an exercise problem for the reader to show that the NAND operation too is nonassociative. Thus, the NOR and NAND gates for more than two inputs must be defined.

The following diagrams serve this purpose, and may be easily extended to any finite number of inputs:

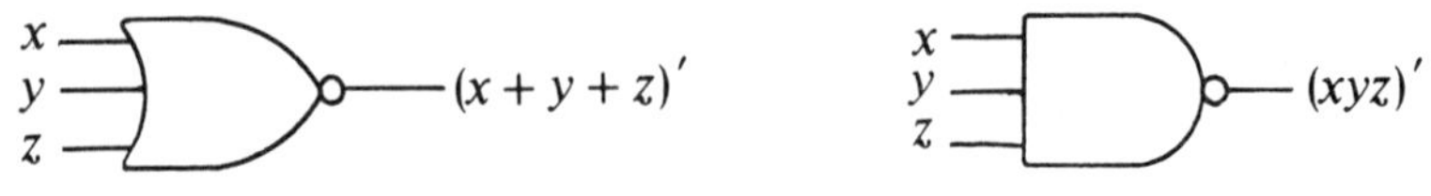

Figure 6.

EXERCISE 4.3

1. Find the truth table for each of the following:
   (a) NAND: $F = (xy)'$
   (b) NOR: $F = x \downarrow y = (x + y)'$
   (c) XOR: $F = x \oplus y = xy' + x'y$
   (d) EQ: $F = x \leftrightarrow y = xy + x'y'$.
2. Implement $((xy)'z)'$ and $(x(yz)')'$ by using NAND gates and verify that NAND is nonassociative.
3. Prove that EQ$(x \leftrightarrow y)$ is associative.
4. Prove that XOR is associative.

In each of Problems 5 through 12, implement the given Boolean function by using logic gates:

5. $(x + y' + z')(y + z' + w')(x' + y' + z')(x' + y + z)$
6. $(xy' + x'y)(zw + z'w')$
7. $x + z + yw + y'w'$
8. $x + zw + z'w'$
9. $x + y'z + yz'w + zw' + y'w'$
10. $x'zw' + y'z'w'$
11. $x + y'w' + yz' + z'w'$
12. $x + yz' + yw' + y'z + zw'$

4. APPLICATIONS TO DIGITAL COMPUTER CIRCUITS

Digital computers perform, among other things, all kinds of arithmetic operations. The most basic arithmetic operation is the addition of two binary digits, which consists of: $0 + 0 = 0$, $0 + 1 = 1$, $1 + 0 = 1$, and $1 + 1 = 10$. Each of the first three operations results in a sum which has a single digit, but the sum in the fourth operation has two digits. The higher bit of this sum is called a *carry*. In adding two multiple digit numbers, a carry is to be added to the next higher digit. The digital circuit that performs the addition of two bits is called a *half-adder*, while the circuit that performs the addition of three bits (two bits plus a previous carry) is called a *full-adder*.

It is now clear that a half-adder needs two binary inputs (denoted by x and y) and two binary outputs denoted by S, for the sum, and C for the carry. The truth table that characterized the performance of the half-adder can be set up as follows:

Table 4

x	y	C	S
0	0	0	0
0	1	0	1
1	0	0	1
1	1	1	0

The fourth row of the truth table, for example, signifies $1 + 1 = 10$. It is easy to obtain the Boolean functions for Table 4. They are:

$$S = x'y + xy' \qquad \text{and} \qquad C = xy$$

Figure 7 shows various ways of implementing these two Boolean functions S and C.

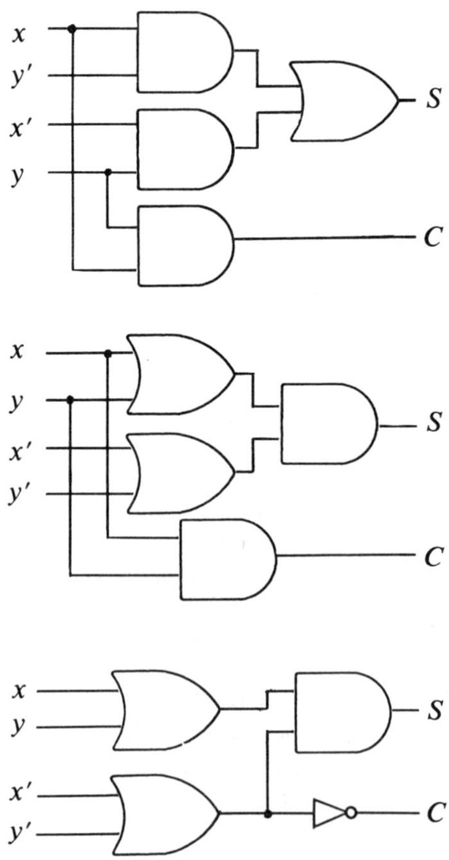

(a) $S = xy' + x'y$
$C = xy$

(b) $S = (x + y)(x' + y')$
$C = xy$

(c) $S = (x + y)(x' + y')$
$C = (x' + y')'$

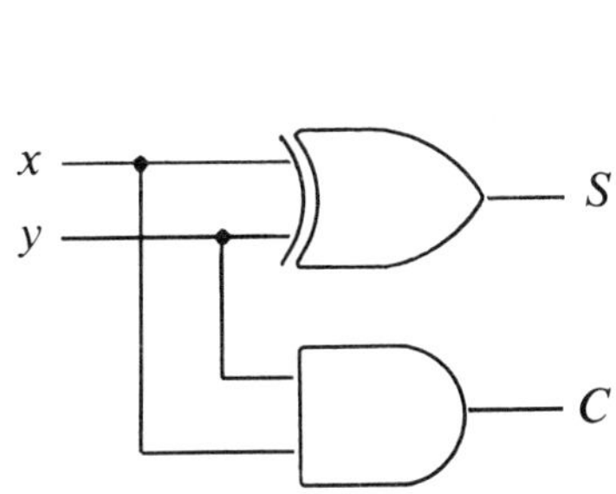

(d) $S = x \oplus y$
$C = xy$

Figure 7

The implementations of the half-adder shown in Figure 7 illustrate that many different designs exist for even such a simple thing as a half-adder. It is a challenge for a computer designer to come up with one that uses the least number of circuit elements.

In a full-adder, two of the input variables (denoted by x and y) represent the two bits to be added. The third variable (denoted by z) represents the carry from the previous lower position. The two outputs, as in the case of the half-adder, denoted by S and C are respectively the sum and carry. We are now ready to write the truth table of the full-adder:

Table 5

x	y	z	C	S
0	0	0	0	0
0	0	1	0	1
0	1	0	0	1
0	1	1	1	0
1	0	0	0	1
1	0	1	1	0
1	1	0	1	0
1	1	1	1	1

From the above truth table, Boolean functions for the full-adder can be directely identified (Cf. Prob. 20, Ex. 4.2):

$$S = xyz + xy'z' + x'yz' + x'y'z$$
$$C = xyz + xyz' + xy'z + x'yz$$

We leave it to the reader to reduce the literals in S and C to:

$$S = (x \oplus y) \oplus z \quad \text{and} \quad C = xy + xz + yz$$

These two functions are directly implemented in Figure 8. Six logic gates are used in this implementation.

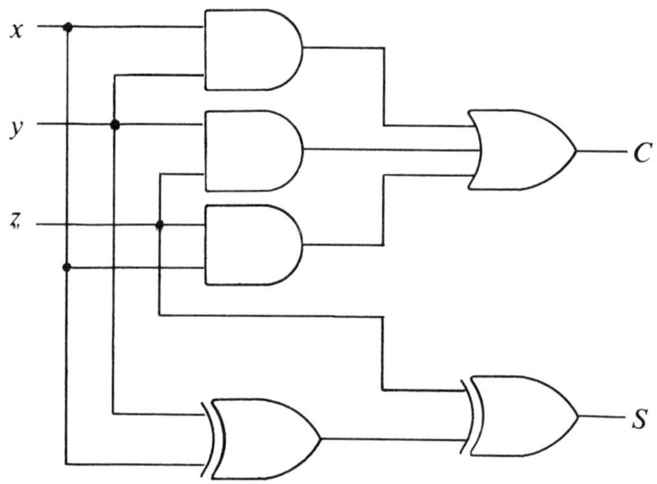

Figure 8

It is, again, a challenging task for a designer to reduce the number of logic gates and yet maintain the identical performance. A fairly good designer should be able to rewrite

$$C = xy + (xy' + x'y)\,z$$

which in conjunction with $S = (x \oplus y) \oplus z$ gives the circuit diagram of Figure 9.

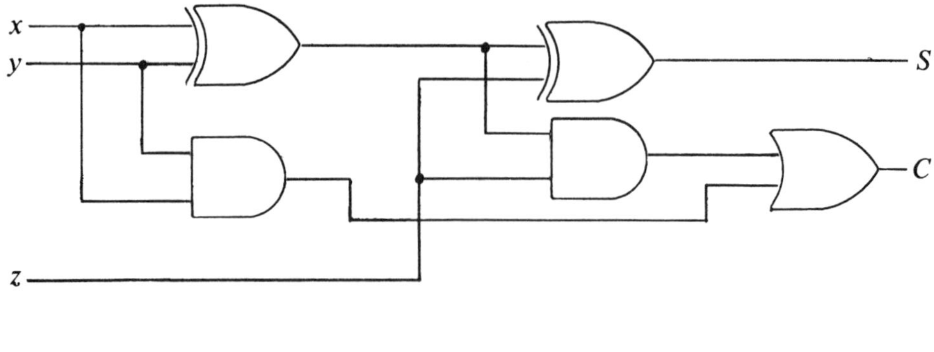

Figure 9

It should be observed that the full-adder of Figure 9 combines two half-adders of Figure 7(d) by an OR gate. Therefore, one may consider one full-adder as good as two half-adders.

EXERCISE 4.4

1. Prove that the following Boolean functions in Figure 7 are equivalent, that is, their truth tables are identical.
   (a) $S = xy' + x'y$
   (b) $S = (x + y)(x' + y')$
   (c) $S = x \oplus y$.
2. Show that $C = xy$ in Figure 7(a) and $C = (x' + y')'$ in Figure 7(c) have the same truth table.
3. Show that, in the full-adder,
$$S = xyz + xy'z' + x'yz' + x'y'z = (x \oplus y) \oplus z.$$
4. Show that
$$\begin{aligned} C &= xyz + xyz' + xy'z + x'yz \\ &= xy + xz + yz \\ &= xy + (xy' + x'y)\,z \end{aligned}$$
5. The performance of a *half-subtractor* is given by the truth table:

x	y	B	D
0	0	0	0
0	1	1	1
1	0	0	1
1	1	0	0

   where D denotes the difference, $x - y$, and B denotes the borrow from the next higher bit.
   (a) Find a set of Boolean functions for B and D.
   (b) Simplify the Boolean functions obtained in (a), if possible.
6. Draw circuit diagrams for (a) and (b) in Problem 5, above.
7. The performance of a *full-subtractor* is given by the truth table:

x	y	z	B	D
0	0	0	0	0
0	0	1	1	1
0	1	0	1	1
0	1	1	1	0
1	0	0	0	1
1	0	1	0	0
1	1	0	0	0
1	1	1	1	1

where x, y, B, D are the same as in the half-subtractor, and z denotes the previous borrow.

(a) Find a set of Boolean functions for D and B.

(b) Simplify the Boolean functions obtained in (a).

8. Draw the circuit diagrams for 7(a) and 7(b), above.
9. Design a full-subtractor that combines two half-subtractors by an OR gate.

Chapter Five

Denumerable Sets and Nondenumerable Sets

Dedekind's definition of an infinite set is used to discuss properties of infinite sets and of finite sets. It is proved, among other things, that denumerable sets are the smallest in "size" among the infinite sets. Properties and examples of denumerable sets and of nondenumerable sets are given.

1. FINITE AND INFINITE SETS

In Section 1, Chapter 2, we mentioned casually that a *finite* set is a set which contains only finitely many elements; although this concept may be developed into a more precise mathematical definition, we prefer an alternate definition (Definition 1) originated by Dedekind.

It was pointed out in Section 1 of Chapter 2 that the set **N** of all natural numbers is an *infinite set*. Let $\mathbf{N}_e = \{2, 4, 6, \ldots\}$ be the set of all even natural numbers. As the reader has shown in Problem 8, Exercise 3.6, there is a one-to-one correspondence between the set **N** and its proper subset $\mathbf{N}_e$.

In other words,

A part is as numerous as the whole.[1]

[1] A striking difference from Euclid's axiom: "The whole is greater than any of its parts" (325 B.C.)

This strange property (of an infinite set) bothered many mathematicians including Georg Cantor. It was Richard Dedekind (1831-1916)[2] who made this property the defining characteristic of an infinite set. The following definition was given by Dedekind in 1888.

Definition 1. A set X is *infinite* provided that it has a proper subset Y such that there exists a one-to-one correspondence between X and Y. A set is *finite* if it is not infinite.

In other words, a set X is infinite if and only if there exists an injection $f: X \to X$ such that $f(X)$ is a proper subset of X. Thus, the set $\mathbf{N}$ of natural numbers is an infinite set.

EXAMPLE 1. The empty set $\varnothing$ and the singleton sets[3] are finite.

Solution. (a) Since the empty set has no proper subset, it cannot be infinite. Therefore, the empty set is finite. (b) Let $\{a\}$ be any singleton set. Since the only proper subset of $\{a\}$ is the empty set $\varnothing$ and there is no one-to-one correspondence between $\{a\}$ and $\varnothing$, $\{a\}$ must be finite.

Theorem 1.

(a) Every superset of an infinite set is infinite.
(b) Every subset of a finite set is finite.

---

[2] Richard Dedekind, one of the greatest mathematicians, was born on October 6, 1831, in Brunswick, Germany. At first, Dedekind's interest lay in physics and chemistry; he considered mathematics merely as the servant of the sciences. But this did not coutinue long; by the age of seventeen he had turned from physics and chemistry to mathematics, whose logic he found more satisfactory. At the age of nineteen he enrolled in the University of Göttingen to study mathematics, and he received his doctor's degree three years later under Gauss. His fundamental contributions to mathematics include the famous "Dedekind Cut," an important concept in the study of irrational numbers, which the reader may have an opportunity to study in a real analysis course.

[3] A *singleton set* is a set which consists of one element alone.

Proof. (a) Let X be an infinite set and let Y be a superset of X, i.e., $X \subseteq Y$. Then by Definition 1 there exists an injection $f: X \to X$ such that $f(X) \neq X$. Define a finction $g: Y \to Y$ by

$$g(y) = \begin{cases} f(y) & \text{if } y \in X \\ y & \text{if } y \in Y - X \end{cases}$$

We leave it to the reader to verify that the function $g: Y \to Y$ is injective and that $g(Y) \neq Y$. It now follows by Definition 1 that Y is infinite.

(b) Let Y be a finite set and let X be a subset of Y, i.e., $X \subseteq Y$. To show that X is finite, we suppose the contrary, that X is infinite. Then by (a), the set Y must be infinite. This is a contradiction. Therefore, the set X is finite.

Theorem 2. Let $g: X \to Y$ be a one-to-one correspondence. If the set X is infinite, then Y is infinite.

Proof. Since X is infinite, by Definition 1 there exists an injection $f: X \to X$ such that $f(X) \neq X$. Since $g: X \to Y$ is a one-to-one correspondence so is $g^{-1}: Y \to X$ (Theorem 14, Chapter 3). We now have the following diagram of injections:

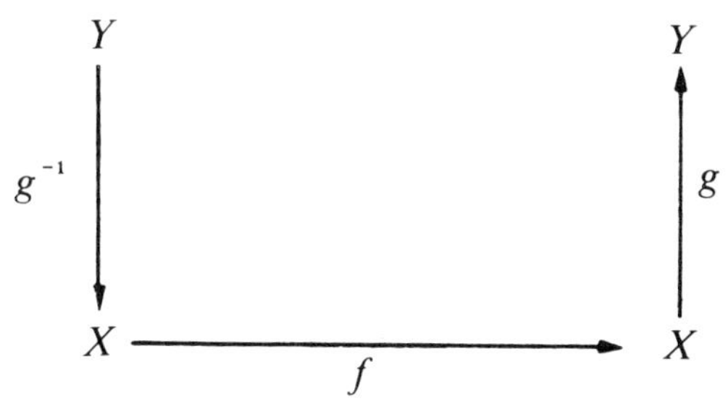

Consequently, the composition $h = g \circ f \circ g^{-1}: Y \to Y$ of injections is an injection [Problem 7, Exercise 3.7]. Finally, we have

$$\begin{aligned} h(Y) &= (g \circ f \circ g^{-1})(Y) = (g \circ f)(g^{-1}(Y)) \\ &= (g \circ f)(X) = g(f(X)) \end{aligned}$$

and $g(f(X)) \neq Y$, because $f(X) \neq X$.

Thus, $h(Y)$ is a proper subset of Y and hence Y is infinite.

Corollary. Let $g: X \to Y$ be a one-to-one correspondence. If the set X is finite, then Y is finite.

Proof. Exercise.

Theorem 3. Let X be an infinite set and let $x_0 \in X$. Then $X - \{x_0\}$ is infinite.

Proof. By Definition 1, there exists an injection $f: X \to X$ such that $f(X) \subset X$. There are two cases to be considered: (1) $x_0 \in f(X)$ or (2) $x_0 \in X - f(X)$. In each case we must construct an injection $g: X - \{x_0\} \to X - \{x_0\}$ such that $g(X - \{x_0\}) \neq X - \{x_0\}$.

Case 1. $x_0 \in f(X)$.
There exists an element x_1 in X such that $f(x_1) = x_0$. A function

$$g: X - \{x_0\} \to X - \{x_0\}$$

may now be defined by

$$g(x) = \begin{cases} f(x) & \text{if } x \neq x_1 \\ x_2 & \text{if } x = x_1 \in X - \{x_0\} \end{cases}$$

where x_2 is an arbitrary fixed element in the nonempty set $X - f(X)$. It follows that $g: X - \{x_0\} \to X - \{x_0\}$ is injective and that $g(X - \{x_0\}) = f(X - \{x_0, x_1\}) \cup \{x_2\} \neq X - \{x_0\}$. Hence $X - \{x_0\}$ is infinite in this case.

Case 2. $x_0 \in X - f(X)$.
Define a function $g: X - \{x_0\} \to X - \{x_0\}$ by $g(x) = f(x)$ for all $x \in X - \{x_0\}$. Since $f: X \to X$ is injective, so is $g: X - \{x_0\} \to X - \{x_0\}$. Finally,

$$g(X - \{x_0\}) = f(X) - \{f(x_0)\} \neq X - \{x_0\}$$

Therefore, in either case, $X - \{x_0\}$ is infinite.

In what follows, let us denote by $\mathbf{N}_k$, $k \in \mathbf{N}$, the set of all natural numbers from 1 to k; that is, $\mathbf{N}_k = \{1, 2, 3, \ldots, k\}$. As an application of Theorem 3, we show the following example that each $\mathbf{N}_k$ is a finite set.

EXAMPLE 2. For each $k \in \mathbf{N}$, the set $\mathbf{N}_k$ is finite.

Proof. We shall prove this by the principle of mathematical induction. By Example 1, the assertion is true for $k = 1$. Now assume that the set $\mathbf{N}_k$ is finite for some natural number k. Consider the set $\mathbf{N}_{k+1} = \mathbf{N}_k \cup \{k + 1\}$. If $\mathbf{N}_{k+1}$ is an infinite set, then by Theorem 3 $\mathbf{N}_{k+1} - \{k + 1\} = \mathbf{N}_k$ is an infinite set, which contradicts the induction hypothesis. Thus, if $\mathbf{N}_k$ is finite then $\mathbf{N}_{k+1}$ is finite. Therefore, by the principle of mathematical induction, the set $\mathbf{N}_k$ is finite for every $k \in \mathbf{N}$.

In effect, there is a close connection between a nonempty finite set and a set $\mathbf{N}_k$.

Theorem 4. A set X is finite if and only if either $X = \varnothing$ or X is in one-to-one correspondence with some $\mathbf{N}_k$.

Proof. If X is either empty or in one-to-one correspondence with some $\mathbf{N}_k$, then by the corollary to Theorem 2 and Examples 1 and 2, the set X is finite.

To show the converse, we show, equivalently, its contrapositive: If $X \neq \varnothing$ and X is not in one-to-one correspondence with any $\mathbf{N}_k$, then X is infinite. Then we can take an element x_1 from X and, again, $X - \{x_1\}$ is not empty; for otherwise we would have $X = \{x_1\}$ in one-to-one correspondence with $\mathbf{N}_1$, a contradiction to the assumption about X. Similarly, we can choose an element x_2 of $X - \{x_1\}$.

Continuing in this manner, suppose that we have chosen elements $x_1, x_2, \ldots, x_k$ of X. Then $X - \{x_1, x_2, \ldots, x_k\}$ is nonempty; otherwise $X = \{x_1, x_2, \ldots, x_k\}$ would be in one-to-one correspondence with $\mathbf{N}_k$, a contradiction to our assumption about X. Thus, we can always choose an element x_{k+1} from $X - \{x_1, x_2, \ldots, x_k\}$. Then, by mathematical induction, for *every* natural number n, there exists a proper subset $\{x_1, x_2, \ldots, x_n\}$ of X. Denote the set of x_n chosen for every natural number n by Y.[4] Then the function f: $Y \to Y - \{x_1\}$ defined by $f(x_k) = x_{k+1}$, for all $k \in \mathbf{N}$, establishes a one-to-one correspondence between Y and its proper subset $Y - \{x_1\}$. Therefore, by Definition 1, Y is infinite and hence, by Theorem 1, X is infinite.

We mention here that Theorem 4 suggests an alternate definition of finite and infinite sets. We can define a set to be finite if and only if it is either an empty set or in one-to-one correspondence with some $\mathbf{N}_k$, and to be infinite if and only if it is not finite. From this alternate definition, our Definition 1 can be proved as a theorem. However, this would require about the same amount of work as our present approach.

---

[4] Here the authors have implicitly used the "axiom of choice," an important axiom to be discussed in Chapter 7. One form of the axiom of choice may be stated as: "Let $\mathscr{P}$ be a nonempty set of nonempty subsets of a given set X. Then there exists a set $R \subseteq X$ such that for every $C \in \mathscr{P}$, $C \cap R$ is a singleton set." This axiom will be used throughout this book without being mentioned explicitly.

EXERCISE 5.1

1. Complete the proof of Theorem 1.
2. Let $g : X \to Y$ be a one-to-one correspondence. Prove that if X is finite, then Y is finite.
3. Prove that the sets **Z**, **Q**, and **R** are infinite.
4. Prove that if A is an infinite set, then so is $A \times A$.
5. Prove that if A and B are infinite sets, then $A \cup B$ is an infinite set.
6. Prove that the union of finitely many finite sets is a finite set.
7. Let A and B be two sets such that $A \cup B$ is infinite. Prove that at least one of the two sets A and B is infinite.
8. Prove the following generalization of Theorem 3: If Y is a finite subset of the infinite set X, then $X - Y$ is infinite.
9. Prove that if A is a set such that every proper superset of A is infinite, then A is infinite.
10. Prove that if B is a set such that every proper subset of B is finite, then B is finite.
11. Prove that if $k \in \mathbf{N}$ and $A_1, A_2, \ldots, A_k$ are sets such that $A_1 \cup A_2 \cup \cdots \cup A_k$ is infinite, then there exists $j \in \mathbf{N}_k$ such that A_j is infinite.
12. Prove that if A is a set such that $A \times A$ is infinite then A is infinite.
13. Prove that if A and B are sets such that $A \times B$ is infinite, then either A or B is infinite.
14. Prove that if $k \in \mathbf{N}$ and $A_1, A_2, \ldots, A_k$ are sets such that $A_1 \times A_2 \times \cdots \times A_k$ is infinite, then there exists $j \in \mathbf{N}_k$ such that A_j is infinite.
15. Prove that if A and B are sets such that $A \oplus B$ is infinite, then either A or B is infinite.
16. Show that the set $\{1, 4, 9, 16, 25, \ldots\}$ is infinite.
17. Prove that the power set of a finite set is finite.

2. EQUIPOTENCE OF SETS

Two finite sets X and Y have the same number of elements if and only if there exists a one-to-one correspondence $f : X \to Y$. Although the phrase "same number of elements" does not apply here if X and Y are infinite, it seems natural to think that two (infinite) sets that are in one-to-one correspondence are of the same size. We formalize this intuition as follows:

Definition 2. Two sets X and Y are said to be *equipotent*, symbolized as $X \sim Y$, provided that there exists a one-to-one correspondence $f : X \to Y$.

Obviously, every set is equipotent to itself. Since the inverse of a one-to-one correspondence is a one-to-one correspondence (Theorem 14, Chapter 3), $X \sim Y$ if and only if $Y \sim X$. *Let us agree that the symbol* $f: X \sim Y$ *represents* "$f: X \to Y$ is a one-to-one correspondence and hence $X \sim Y$." Using this convenient notation, the first half of the result of Problem 12, Exercise 3.7 may be restated as: If $f: X \sim Y$ and $g: Y \sim Z$ then $g \circ f: X \sim Z$. We have thus proved the following theorem.

Theorem 5. Let $\mathcal{S}$ be a set of sets and let $\mathcal{R}$ be a relation on $\mathcal{S}$ given by $X \mathcal{R} Y$ if and only if X and Y are members of $\mathcal{S}$ and $X \sim Y$. Then $\mathcal{R}$ is an equivalence relation on $\mathcal{S}$.

In the following example the symbols (0, 1) and (−1, 1) represent open intervals of real numbers, not ordered pairs of integers.

EXAMPLE 3.

(a) $(0, 1) \sim (-1, 1)$.

(b) $(-1, 1) \sim \mathbf{R}$ and $(0, 1) \sim \mathbf{R}$.

Solution. (a) The function $f: (0, 1) \to (-1, 1)$ given by $f(x) = 2x - 1$ is a one-to-one correspondence. Hence $(0, 1) \sim (-1, 1)$.

(b) The trigonometric function $g: (-1, 1) \to \mathbf{R}$ given by $g(x) = \tan(\pi x/2)$ is a one-to-one correspondence; therefore $(-1, 1) \sim \mathbf{R}$. The reader should verify this assertion by sketching a graph of $g(x) = \tan(\pi x/2)$. A rigorous proof may be obtained by verifying the following two observations:

(1) $g: (-1, 1) \to \mathbf{R}$ is continous and unbounded both below and above.

(2) $g'(x) = (\pi/2)\sec^2(\pi x/2) > 0 \ \forall \ x \Rightarrow g$ is strictly increasing.

Since the equipotence "relation" $\sim$ is transitive,[5] $(0, 1) \sim (-1, 1)$ and $(-1, 1) \sim \mathbf{R}$ imply $(0, 1) \sim \mathbf{R}$.

[5] Strictly speaking, "$\sim$" is not a relation, because its domain is not a set (see Theorem 10 of Chapter 2). But, we may call it a relation here if we consider it to be defined on any given set of sets $\mathcal{S}$ (Theorem 5).

Theorem 6. Let X, Y, Z, and W be sets with $X \cap Z = \varnothing = Y \cap W$, and let $f: X \sim Y$ and $g: Z \sim W$. Then $f \cup g: (X \cup Z) \sim (Y \cup W)$.

Proof. Since $f: X \rightarrow Y$ and $g: Z \rightarrow W$ are functions with $X \cap Z = \varnothing$, by Theorem 8 of Chapter 3, $f \cup g: X \cup Z \rightarrow Y \cup W$ is a function. We leave it to the reader to prove that the latter function is a one-to-one correspondence.

Theorem 7. Let X, Y, Z, and W be sets such that $X \sim Y$ and $Z \sim W$. Then $X \times Z \sim Y \times W$.

Proof. Let $f: X \sim Y$ and $g: Z \sim W$. We define the function $f \times g: X \times Z \rightarrow Y \times W$ by $(f \times g)(x, z) = (f(x), g(z))$ for all $(x, z) \in X \times Z$. We ask the reader to verify that the latter function is a one-to-one correspondence.

Examining the various finite sets $\mathbf{N}_k = \{1, 2, 3, \ldots, k\}$ as k increases and noting that the infinite sets $\mathbf{Z}$, $\mathbf{Q}$, and $\mathbf{R}$ (see Problem 3, Exercise 5.1) are supersets of $\mathbf{N}$, it appears that the "smallest" infinite set is the set $\mathbf{N}$ of all natural numbers, or any set that is equipotent to $\mathbf{N}$. We shall soon learn, in Section 4, that not all infinite sets are equipontent to $\mathbf{N}$.

Definition 3. A set X is said to be *denumerable* provided that $X \sim \mathbf{N}$. A *countable* set is a set which is either finite or denumerable.

Let X be a denumerable set. Then there is a one-to-one correspondence $f: \mathbf{N} \sim X$. If we denote

$$f(1) = x_1,\ f(2) = x_2,\ f(3) = x_3, \ldots,\ f(k) = x_k, \ldots$$

then X may be alternatively denoted as $\{x_1, x_2, x_3, \ldots, x_k, \ldots\}$; the dots are used to indicate that the elements are labeled in a definite order as indicated by the subscripts. An explanation of the term "countable" is now in order. For a finite set, it is theoretically possible to count its elements and the term is suitable. Even though the actual counting of all elements of a denumerable set $X = \{x_1, x_2, x_3, \ldots\}$ is impossible, nevertheless, the set X is in one-to-one correspondence with the natural, or counting, numbers.

Theorem 8. Every infinite subset of a denumerable set is denumerable.

Proof. Let Y be an infinite subset of the denumerable set $X = \{x_1, x_2, x_3, \ldots\}$.

Let n_1 be the smallest subscript for which $x_{n_1} \in Y$, and let n_2 be the smallest subscript for which $x_{n_2} \in Y - \{x_{n_1}\}$. Having defined $x_{n_{k-1}} \in Y$, let n_k be the smallest subscript such that $x_{n_k} \in Y - \{x_{n_1}, x_{n_2}, \ldots, x_{n_{k-1}}\}$. Such an x_{n_k} always exists for each $k \in \mathbf{N}$, because Y is infinite, which ensures that $Y - \{x_{n_1}, x_{n_2}, \ldots, x_{n_{k-1}}\} \neq \varnothing$ for each $k \in \mathbf{N}$. We have thus constructed a one-to-one correspondence $f: \mathbf{N} \sim Y$ where $f(k) = x_{n_k}$ for each $k \in \mathbf{N}$. Therefore, Y is denumerable.

A shorter but less intuitive alternate proof of Theorem 8 is indicated in Problem 10 at the end of this section. The following corollary is an immediate consequence of Definition 3 and Theorem 8.

Corollary. Every subset of a countable set is countable.

More examples and properties of denumerable sets are give in the next section.

EXERCISE 5.2

1. Complete the proof of Theorem 6.
2. Complete the proof of Theorem 7.
3. Prove that if X and Y are two sets, then $X \times Y \sim Y \times X$.
4. Prove that if $(X - Y) \sim (Y - X)$ then $X \sim Y$.
5. Prove the following generalization of Theorem 6: Let $\{X_\gamma \mid \gamma \in \Gamma\}$ and $\{Y_\gamma \mid \gamma \in \Gamma\}$ be two families of disjoint sets such that $X_\gamma \sim Y_\gamma$ for each $\gamma \in \Gamma$. Then $\bigcup_{\gamma \in \Gamma} X_\gamma \sim \bigcup_{\gamma \in \Gamma} Y_\gamma$.
6. Prove that if X is a denumerable set and Y is a finite subset of X, then $X - Y$ is denumerable. [Compare with Problem 8, Exercise 5.1]
7. Prove that if X is a denumerable set and Y is a finite set, then $X \cup Y$ is denumerable.
8. Prove that the set $\mathbf{N}_e$ of all even natural numbers and the set $\mathbf{N}_o$ of all odd natural numbers are denumerable.
9. Let A be a nonempty set, and let 2^A be the set of all functions from the set A to the set $\{0, 1\}$. Prove that $\mathscr{P}(A) \sim 2^A$.
10. Let X be a denumerable set and Y an infinite subset of X. Let $g: X \sim \mathbf{N}$, and let $h: Y \to \mathbf{N}$ be defined by

    $$h(y) = \text{the number of elements in } \{1, 2, 3, \ldots, g(y)\} \cap g(Y)$$

    Prove that h is a one-to-one correspondence and hence that Y is denumerable.
11. Let $[a, b]$, $a < b$, be any closed interval. Prove that the function $f: [0, 1] \to [a, b]$ defined by $f(x) = a + (b - a)x$, for all x in $[0, 1]$,

establishes the equipotence: $[0, 1] \sim [a, b]$. Thus, $[a, b] \sim [c, d]$ for any real numbers c and d such that $c < d$.

12. Let a, b, c, and d be any real numbers such that $a < b$ and $c < d$. Prove that:
    (a) $[0, 1] \sim (0, 1]$
    (b) $[a, b] \sim (c, d]$
    (c) $(c, d] \sim [c, d)$.
13. Prove, under the assumptions of Problem 12, that:
    (a) $[0, 1] \sim (0, 1)$
    (b) $[a, b] \sim (c, d)$
    (c) $(a, b] \sim (c, d]$.

3. EXAMPLES AND PROPERTIES OF DENUMERABLE SETS

The set $\mathbf{N}_e$ of all even natural numbers and the set $\mathbf{N}_o$ of all odd natural numbers are denumerable (Problem 8, Exercise 5.2). Since the union $\mathbf{N}_e \cup \mathbf{N}_o$ $(= \mathbf{N})$ of these two denumerable sets is denumerable, the next theorem should be predictable.

Theorem 9. The union of two denumerable sets is denumerable.

Proof. Let A and B be any two denumerable sets. We shall show that $A \cup B$ is denumerable in the following two cases:

Case 1. $A \cap B = \varnothing$.

Since $A \sim \mathbf{N}$ and $\mathbf{N} \sim \mathbf{N}_o$, we have $A \sim \mathbf{N}_o$. Similarly, we have $B \sim \mathbf{N}_e$. Consequently, by Theorem 6, we have $(A \cup B) \sim (\mathbf{N}_o \cup \mathbf{N}_e) = \mathbf{N}$, which shows that $A \cup B$ is denumerable.

Case 2. $A \cap B \neq \varnothing$.

Let $C = B - A$. Then $A \cup C = A \cup B$ and $A \cap C = \varnothing$; the set $C \subseteq B$ is either finite of denumerable [corollary to Theorem 8]. If C is finite, by Problem 7 of Exercise 5.2, $A \cup C$ is denumerable, whereas if C is denumerable then $A \cup C$ is denumerable by case 1, above.

Therefore, the set $A \cup B$ is denumerable.

Corollary. Let $A_1, A_2, \ldots, A_n$ be denumerable sets. Then $\bigcup_{k=1}^{n} A_k$ is denumerable.

Proof. Left to the reader as an exercise.

We ask the reader to verify the next example.

EXAMPLE 4. The set **Z** of all integers is denumerable.

Theorem 10. The set $\mathbf{N} \times \mathbf{N}$ is denumerable

Proof. Consider the function $f: \mathbf{N} \times \mathbf{N} \to \mathbf{N}$ given by

$$f(j, k) = 2^j 3^k$$

for all $(j, k) \in \mathbf{N} \times \mathbf{N}$. This function is injective, so that

$$\mathbf{N} \times \mathbf{N} \sim f(\mathbf{N} \times \mathbf{N}) \subseteq \mathbf{N}.$$

Since $\mathbf{N} \times \mathbf{N}$ is infinite, so is $f(\mathbf{N} \times \mathbf{N})$. By Theorem 8, $f(\mathbf{N} \times \mathbf{N})$ is denumerable and hence the set $\mathbf{N} \times \mathbf{N}$ is denumerable.

Corollary. For each $k \in \mathbf{N}$ let A_k be a denumerable set satisfying $A_j \cap A_k = \varnothing$ for all $j \neq k$. Then $\bigcup_{k \in \mathbf{N}} A_k$ is denumerable.[6]

Proof. For each $k \in \mathbf{N}$, let $f_k : \mathbf{N} \to \mathbf{N} \times \{k\}$ be the function given by $f_k(j) = (j, k)$ for all $j \in \mathbf{N}$. Clearly, each $f_k : \mathbf{N} \to \mathbf{N} \times \{k\}$ is a one-to-one correspondence. That is, $\mathbf{N} \sim \mathbf{N} \times \{k\}$. Since $A_k \sim \mathbf{N}$ and $\mathbf{N} \sim \mathbf{N} \times \{k\}$ for each $k \in \mathbf{N}$, we have $A_k \sim \mathbf{N} \times \{k\}$ for each $k \in \mathbf{N}$. It then follows from Problem 5 of Exercise 5.2 that $\bigcup_{k \in \mathbf{N}} A_k \sim \bigcup_{k \in \mathbf{N}} \mathbf{N} \times \{k\}$. But the set $\bigcup_{k \in \mathbf{N}} \mathbf{N} \times \{k\}$ equals the denumerable set $\mathbf{N} \times \mathbf{N}$. Therefore, $\bigcup_{k \in \mathbf{N}} A_k$ is denumerable.

EXAMPLE 5. The set **Q** of all rational numbers is denumerable.

Proof. We shall represent each rational number uniquely as p/q, where $p \in \mathbf{Z}$, $q \in \mathbf{N}$ and the greatest common divisor of p and q is 1. Let $\mathbf{Q}_+$ be the set of all such $p/q > 0$, and let $\mathbf{Q}_- = \{-p/q \mid p/q \in \mathbf{Q}_+\}$. Then $\mathbf{Q} = \mathbf{Q}_+ \cup \{0\} \cup \mathbf{Q}_-$. It is evident that $\mathbf{Q}_+ \sim \mathbf{Q}_-$. Hence, to show that $\mathbf{Q}$ is denumerable, it is sufficient to show that $\mathbf{Q}_+$ is denumerable. To this end, we consider the function $f: \mathbf{Q}_+ \to \mathbf{N} \times \mathbf{N}$ given by $f(p/q) = (p, q)$. Since this function is injective, we have $\mathbf{Q}_+ \sim f(\mathbf{Q}_+) \subseteq \mathbf{N} \times \mathbf{N}$. Since $\mathbf{Q}_+$, as a superset of $\mathbf{N}$, is infinite, $f(\mathbf{Q}_+)$ is an infinite subset of the denumerable set

[6] This result is true without the hypothesis "$A \cap A_k = \varnothing$ for all $j \neq k$." See Problem 7.

$\mathbf{N} \times \mathbf{N}$. Therefore, $f(\mathbf{Q}_+)$ is denumerable and consequently $\mathbf{Q}_+$ is denumerable. The proof is now complete.

The next theorem indicates that the denumerable sets are, in a sense, the smallest in "size" among the infinite sets.

Theorem 11. Every infinite set contains a denumerable subset.

Proof. Let X be any given infinite set. Then $X \neq \varnothing$ so that we can pick an element, call it x_1, from the set X. Next, let x_2 be an element in $X - \{x_1\}$. Similarly, pick an element x_3 from the nonempty set $X - \{x_1, x_2\}$. Having so defined x_{k-1} we let x_k be an element of $X - \{x_1, x_2, \ldots, x_{k-1}\}$. Such an x_k exists for every $k \in \mathbf{N}$, because X is infinite, which ensures that $X - \{x_1, x_2, \ldots, x_{k-1}\} \neq \varnothing$ for every $k \in \mathbf{N}$. The set $\{x_k \mid k \in \mathbf{N}\}$ is a denumerable subset of X, and the proof is now complete.

EXERCISE 5.3

1. Prove the assertion of Example 4: The set $\mathbf{Z}$ of all integers is denumerable
2. Prove the corollary to Theorem 9.
3. Prove that the union of finitely many countable sets is countable.
4. Prove that if A and B are denumerable sets, then so is $A \times B$. In particular $\mathbf{Z} \times \mathbf{N}$, $\mathbf{Z} \times \mathbf{Z}$, and $\mathbf{Q} \times \mathbf{Q}$ are denumerable.
5. Find an injection $f: \mathbf{Q} \to \mathbf{Z} \times \mathbf{N}$ and give an alternate proof for Example 5.
6. Prove that the set of all circles in the Cartesian plane having rational radii and centers at points with both coordinates rational is denumerable.
7. Prove that if for each $k \in \mathbf{N}$, B_k is a denumerable set, then $\bigcup_{k \in \mathbf{N}} B_k$ is denumerable.
8. Prove that if X is a countable set and $f: X \to Y$ is a surjection, then Y is countable.
9. Prove that every denumerable set X has a denumerable subset Y such that $X - Y$ is denumerable.
10. Prove that the set of all polynomials
$$a_0x^n + a_1x^{n-1} + \cdots + a_n$$
with integer coefficients is denumerable. [Hint: Use the Corollary to Theorem 10.]
11. An *algebraic number* is, by definition, any real root of an equation $a_0x^n + a_1x^{n-1} + \cdots + a_n = 0$ with integer coefficients. Prove that the set of all algebraic numbers is denumerable.
12. Prove that the set of all finite subsets of a denumerable set is denumerable.

4. NONDENUMERABLE SETS

All infinite sets that we have seen so far have been denumerable. This may lead the reader to wonder whether all infinite sets are denumerable. It is commonly thought that Georg Cantor tried to prove that every infinite set is denumerable when he first began his development of set theory. However, he surprised himself by proving that there exist nondenumerable sets.

Theorem 12. The open unit interval $(0, 1)$ of real numbers is a nondenumerable set.

Proof. Let us first express each number x, $0 < x < 1$, as a decimal expansion in the form $.x_1x_2x_3\ldots$, where $x_n \in \{0, 1, 2, \ldots, 9\}$ for all $n \in \mathbf{N}$. For example, $1/3 = .333\ldots$, $\sqrt{2}/2 = .707106\ldots$. In order to have a unique infinite decimal expression, for those numbers with a terminating decimal expansion such as $1/4 = .25$, let us agree to decrease the last digit by one and append 9's so that $1/4 = .24999\ldots$ not $.25000\ldots$. Under this agreement, two numbers in the interval $(0, 1)$ are equal if and only if the corresponding digits in their decimal expansions are identical. Thus, if two such numbers $x = .x_1x_2x_3\ldots$ and $y = .y_1y_2y_3\ldots$ have one decimal place, say the kth decimal place, such that $x_k \neq y_k$, then $x \neq y$. This is a crucial point upon which our proof rests.

Now suppose that the set $(0, 1)$ is denumerable. Then there exists a one-to-one correspondence $f: \mathbf{N} \sim (0, 1)$. So we may list all elements of $(0, 1)$ as follows:

$$
(*) \qquad \begin{aligned}
f(1) &= .a_{11}a_{12}a_{13}\ldots \\
f(2) &= .a_{21}a_{22}a_{23}\ldots \\
f(3) &= .a_{31}a_{32}a_{33}\ldots \\
&\vdots \\
f(k) &= .a_{k1}a_{k2}a_{k3}\ldots \\
&\vdots
\end{aligned}
$$

where each $a_{jk} \in \{0, 1, 2, \ldots, 9\}$.

We shall construct a number $z \in (0, 1)$ which cannot be found in the above listing of $f(k)$'s. This contradiction will imply that our earlier supposition that $(0, 1)$ is denumerable was wrong and that the set $(0, 1)$ is nondenumerable. Let $z = .z_1z_2z_3\ldots$ be defined by $z_k = 5$ if $a_{kk} \neq 5$ and $z_k = 1$ if $a_{kk} = 5$, for each $k \in \mathbf{N}$. The number $z = .z_1z_2z_3\ldots$ clearly satisfies $0 < z < 1$; but $z \neq f(1)$ since $z_1 \neq a_{11}$, $z \neq f(2)$ since $z_2 \neq a_{22}, \ldots,$

and in general $z \neq f(k)$ since $z_k \neq a_{kk}$, for all $k \in \mathbf{N}$. Therefore, $z \notin f(\mathbf{N}) = (0, 1)$. We have now the promised contradiction, and the proof is complete.

Corollary. The set **R** of all real numbers is nondenumerable.

Proof. We have proved, in Example 3(b), that $\mathbf{R} \sim (0, 1)$. Now $(0, 1)$ is nondenumerable; therefore its equipotent set **R** must be nondenumerable (see Problem 1).

EXAMPLE 6. The set of all irrational numbers is nondenumerable.

Proof. We have shown, in Example 5, that the set **Q** of all rational numbers is denumerable. The set of all irrational numbers is, by definition, the set $\mathbf{R} - \mathbf{Q}$. It is easy to see that $\mathbf{R} - \mathbf{Q}$ is an infinite set. To show that $\mathbf{R} - \mathbf{Q}$ is nondenumerable, we suppose the contrary, that $\mathbf{R} - \mathbf{Q}$ is denumerable. It then follows that the union $(\mathbf{R} - \mathbf{Q}) \cup \mathbf{Q} = \mathbf{R}$ is denumerable (Theorem 9). This contradicts the corollary to Theorem 12. Therefore the set $\mathbf{R} - \mathbf{Q}$ of all irrational numbers is nondenumerable.

Remarks. (1) The method of proof used in Theorem 12 is called Cantor's diagonal method, because it was originated by Cantor and the construction of the key number $z = .z_1z_2z_3\ldots$ in the proof is based on the digits a_{11}, a_{22}, $a_{33}, \ldots$ on the principal diagonal of the table (*) of digits. This proof, though if might not be easy for the beginner to appreciate, reveals Cantor's ingenuity.

(2) The existence of nondenumerable sets shows that there are classes of infinite sets. In fact, as the reader shall see in the next chapter, there is an abundance of "equipotence classes" of infinite sets.

EXERCISE 5.4

1. Let A and B be two equipotent sets. Prove that if A is nondenumerable, then B is nondenumerable.
2. Prove that every superset of an uncountable set is uncountable.
3. Using the result of Problem 2, above, give an alternate proof for the corollary to Theorem 12.

4. Prove that the set of all irrational numbers between 0 and 1 is nondenumerable.
5. A *transcendental number* is, by definition, a real number that is not algebraic (see Problem 11, Exercise 5.3). Prove that the set of all trancendental numbers is nondenumerable.
6. Let $S^1 = \{(x, y) \in \mathbf{R} \times \mathbf{R} \mid x^2 + y^2 = 1\}$. Prove that $S^1 \sim \mathbf{R}$ and hence nondenumerable.
7. Prove that a set A is nondenumerable if and only if $A \times A$ is nondenumerable.
8. Prove that if A and B are sets such that $A \times B$ is nondenumerable, then either A or B is nondenumerable.
9. Prove that if $k \in \mathbf{N}$ and $A_1, A_2, \ldots, A_k$ are sets such that $A_1 \times A_2 \times \cdots \times A_k$ is nondenumerable, then there exists $j \in \mathbf{N}_k$ such that A_j is nondenumerable.
10. Prove that if $k \in \mathbf{N}$ and $A_1, A_2, \ldots, A_k$ are sets such that $A_1 \cup A_2 \cup \cdots \cup A_k$ is nondenumerable, then there exists $j \in \mathbf{N}_k$ such that A_j is nondenumerable.
11. Prove that if A and B are sets such that $A \oplus B$ is nondenumerable, then either A or B is nondenumerable. Is the converse true?

Chapter Six

Cardinal Numbers and Cardinal Arithmetic

The concept of cardinal numbers is introduced. Similarities and distinctions between properties of finite and transfinite cardinal numbers are exhibited in the course of exploring cardinal arithmetic—addition, multiplication, and exponentiation. The chapter ends with a historical remark on the (generalized) continuum hypothesis.

1. THE CONCEPT OF CARDINAL NUMBERS

Very naturally, the concept of numbers entered our lives early. We were able to notice, for example, the similarity between three apples and three oranges and the distinction between two fingers and four fingers. Although we had a concept of number, most of us did not have a precise definition of number. We have known, for example, that $2 + 3 = 5$, $3 < 4$, $6 \times 7 = 42$, etc. This leads us to believe that we don't need to know what a number really is; what we should know are equality and order between numbers, and how to calculate with numbers—just as chess players are not concerned with what a knight is, but rather with how it performs. Therefore, we do not define here what a *cardinal number* is,[1] but just introduce it as a primitive concept relating to the "size" of sets. The important rules guiding this new concept are

C-1. Each set A is associated with a cardinal number, denoted by card A, and for each cardinal number a there is a set A with card $A = a$.

[1] Perhaps the reader should be informed that it is possible to define the cardinal numbers as the "initial ordinals." See page 183.

C-2. Card$A$ = 0 if and only if $A = \varnothing$.
C-3. If A is a nonempty finite set, i.e., $A \sim \{1, 2, 3, \ldots, k\}$ for some $k \in \mathbf{N}$, then card$A = k$.
C-4. For any two sets A and B, card$A$ = card$B$ if and only if $A \sim B$.

The rules C-2 and C-3 define the cardinal numbers of the finite sets—*the cardinal number of a finite set is the number of elements in that set.* In axiomatic treatments of set theory, C-1 and C-4 are usually postulated as an axiom, called the *axiom of cardinality*. The beginner might find C-1 and C-4 rather difficult to accept, because these rules do not say much about card A when A is an infinite set. This difficulty will be overcome gradually as we proceed—just as the reader did not know what calculus was until he was about halfway through the course. At this stage, we may say, roughly, that the cardinal number of a set is the property that the set has in common with all sets that are equipotent to it.

EXERCISE 6.1

1. Show that the natural numbers are cardinal numbers.
2. Give three cardinal numbers which are not natural numbers.
3. Show that card$\mathbf{N}$ = card$\mathbf{Z}$ = card$\mathbf{Q}$ = card$\mathbf{Q} \times \mathbf{Q}$.
4. Let A be a set and x an element not in A. Prove that if card$(A \cup \{x\})$ = card$A$, then A is infinite.
5. Is the converse of Problem 4, above, true?
6. Prove that if card A = card B = card $\mathbf{N}$, then card $(A \cup B)$ = card B.
7. Let $\mathbf{N}^1 = \mathbf{N}$ and $\mathbf{N}^{n+1} = \mathbf{N}^n \times \mathbf{N}$ for all $n \in \mathbf{N}$. Prove that
   (a) card$\mathbf{N}^k$ = card$\mathbf{N}$ for all $k \in \mathbf{N}$
   (b) card $\bigcup_{k \in \mathbf{N}} \mathbf{N}^k$ = card$\mathbf{N}$.

2. ORDERING OF THE CARDINAL NUMBERS—THE SCHRÖDER—BERNSTEIN THEOREM

We shall call the cardinal number of a finite set a *finite* cardinal number, and the cardinal number of an infinite set a *transfinite* cardinal number. The rules C-2 and C-3 of the previous section show that the finite cardinal numbers are precisely the nonnegative integers. Thus, the finite cardinal numbers have an inherited natural order: $0 < 1 < 2 < \cdots < k < k+1 < \cdots$. For any two transfinite cardinal numbers, the rule C-4 tells us when they are equal and when they are not equal. But we will not be satisfied with just that; when they are unequal we wish to be able to tell which one is "less" than the other.

Definition 1. Let A and B be sets. Then card$A$ is said to be *less than or equal to* card$B$, denoted by card$A \leq$ card$B$, provided that the set A is equipotent to a subset of B. If card$A \leq$ card$B$ and card$A \neq$ card$B$, then we write card$A <$ card$B$.

Although this definition is designed to order transfinite cardinal numbers, it applies to finite cardinal numbers as well, and when it is applied to finite cardinal numbers the result is the same as the traditional natural ordering mentioned above.

EXAMPLE 1. card$\mathbf{N} <$ card$\mathbf{R}$.

Proof. Since the set $\mathbf{N}$ is a subset of $\mathbf{R}$, $\mathbf{N}$ is equipotent to a subset of $\mathbf{R}$, $\mathbf{N} \sim \mathbf{N} \subset \mathbf{R}$, but from Section 4, Chapter 5, we know that the infinite set $\mathbf{R}$ is nondenumerable. Hence, $\mathbf{R}$ is not equipotent to the set $\mathbf{N}$. By Definition 1, we have card$\mathbf{N} <$ card$\mathbf{R}$.

So far it is not clear to us how two cardinal numbers card$A$ and card$B$ compare when the set A is equipotent to a subset of B and the set B is equipotent to a subset of A. Georg Cantor conjectured that, in this case, card$A$ should be equal to card$B$. Later, in the 1890's, this conjecture was proved, independently, both by F. Bernstein in Cantor's seminar, and by E. Schroder on the basis of a logical calculus. This celebrated result is now generally known as the Schröder-Bernstein Theorem.

Theorem 1. (*Schröder-Bernstein Theorem*). If A and B are sets such that A is equipotent to a subset of B and B is equipotent to a subset of A, then A and B are equipotent.

We shall first prove the following special case of Theorem 1, from which Theorem 1 follows easily.

Lemma. If B is a subset of A and if there exists an injection $f: A \to B$, then there is a bijection $h: A \sim B$.

Proof. If B is A, then the identity function on A is such an h. Suppose that B is a proper subset of A, and let C denote the set $\bigcup_{n \geq 0} f^n(A - B)$, where f^0 is the identity function on A and, for each positive integer k and for each $x \in A$,

$f^k(x) = f(f^{k-1}(x))$. For each z in A, define $h(z)$ as follows:

$$h(z) = \begin{cases} f(z) & \text{if } z \in C \\ z & \text{if } z \in A - C \end{cases}$$

Observe that $A - B$ is a subset of $C, f(C) \subseteq C$, and that if m and n are two distinct nonnegative integers, say $m < n$, then $f^m(A - B)$ and $f^n(A - B)$ are disjoint. For otherwise, there exist x and x' in $A - B$ such that $f^m(x) = f^n(x')$, which leads to $f^{n-m}(x') = x \in B \cap (A - B)$, a contradiction. Finally, by the definition of h and the last observation, we have

$$\begin{aligned} h(A) &= (A - C) \cup f(C) \\ &= \left[A - \bigcup_{n \geq 0} f^n(A - B)\right] \cup f\left(\bigcup_{n \geq 0} f^n(A - B)\right) \\ &= \left[A - \bigcup_{n \geq 0} f^n(A - B)\right] \cup \left[\bigcup_{n \geq 1} f^n(A - B)\right] \\ &= A - (A - B) \\ &= B \end{aligned}$$

From these observations and the fact that f is injective, it follows that $h: A \to B$ is a bijection. This completes the proof of the lemma.

The main idea behind the above proof may be visualized in the following diagram, where the whole rectangle represents the set A:

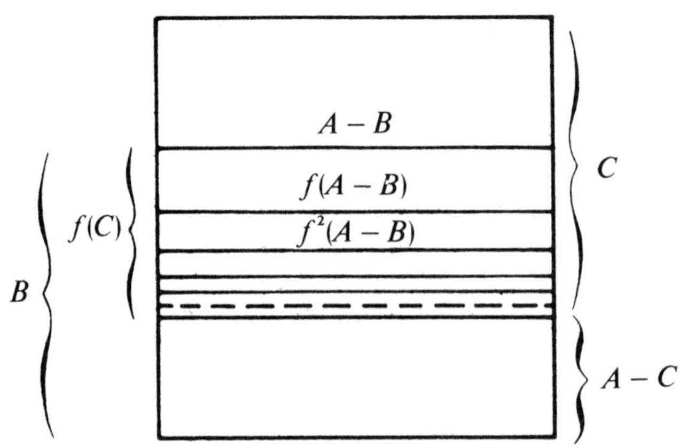

Figure 17.

Proof (Theorem 1). Let A_0 and B_0 be subsets of A and B, respectively, such that $A \sim B_0$ and $B \sim A_0$, and let $f_0 : A \sim B_0$ and $g_0 : B \sim A_0$ be two bijections. Let $f: A \to A_0$ be given by $f(x) = g_0(f_0(x))$, which is an injection. Hence, by the above lemma, there is a bijection $h : A \sim A_0$. Consequently, the composition $g_0^{-1} \circ h : A \sim B$ of two bijections $h : A \sim A_0$ and $g_0^{-1} : A_0 \sim B$ is a bijection.[2]

Corollary. If A and B are sets such that $\operatorname{card} A \leq \operatorname{card} B$ and $\operatorname{card} B \leq \operatorname{card} A$, then $\operatorname{card} A = \operatorname{card} B$.

So far we know very little about transfinite cardinal numbers, because we have seen only two such cardinal numbers, $\operatorname{card} \mathbf{N}$ and $\operatorname{card} \mathbf{R}$. Naturally, we would like to know whether there are other transfinite cardinal numbers. The answer to this question is given in the next section—there is indeed an unlimited supply of distinct transfinite cardinal numbers.

Another important question is this: If m and n are two distinct finite cardinal numbers, then either $m < n$ of $n < m$; is this true for transfinite cardinal numbers? The answer is Yes, but the proof depends on a result of the next chapter and is therefore postponed until Theorem 4 of Chapter 7.

EXERCISE 6.2

1. Let n be any finite cardinal number. Prove that $n < \operatorname{card} \mathbf{N}$.
2. Let a be any transfinite cardinal number. Prove that $\operatorname{card} \mathbf{N} \leq a$. Thus, $\operatorname{card} \mathbf{N}$ is the smallest transfinite cardinal number.
3. Let A and B be sets. Prove that $\operatorname{card} A \leq \operatorname{card} B$ if and only if there exists an injection $f: A \to B$.
4. Let A, B, and C be sets. Prove that
   (a) If $\operatorname{card} A \leq \operatorname{card} B$ and $\operatorname{card} B \leq \operatorname{card} C$, then $\operatorname{card} A \leq \operatorname{card} C$.
   (b) If $\operatorname{card} A < \operatorname{card} B$ and $\operatorname{card} B < \operatorname{card} C$, then $\operatorname{card} A < \operatorname{card} C$.
5. Let A, B, and C be sets. Prove that
   (a) If $\operatorname{card} A \leq \operatorname{card} B$ and $\operatorname{card} B < \operatorname{card} C$ then $\operatorname{card} A < \operatorname{card} C$.
   (b) If $\operatorname{card} A < \operatorname{card} B$ and $\operatorname{card} B \leq \operatorname{card} C$ then $\operatorname{card} A < \operatorname{card} C$.
6. Prove that if A and B are sets such that $A \subseteq B$, then $\operatorname{card} A \leq \operatorname{card} B$.
7. Prove that if A, B, and C are sets such that $A \subseteq B \subseteq C$ and $A \sim C$, then $A \sim B$.
8. Show that $\operatorname{card} \mathbf{N} < \operatorname{card} \mathbf{R}$.
9. Let A be a set, and x be an element not in A. Prove that if $\operatorname{card} A < \operatorname{card}(A \cup \{x\})$, then A is finite.

---

[2] This proof and the preceding lemma are adopted from R. H. Cox, "A Proof of the Schröder-Bernstein Theorem," American Mathematical Monthly, **75**, No. 5 (1968), 508.

3. CARDINAL NUMBER OF A POWER SET—CANTOR'S THEOREM

Let X be a set. Recall that the power set $\mathscr{P}(X)$ of X is the set of all subsets of X (Section 2, Chapter 2). Georg Cantor himself proved that $\operatorname{card} X < \operatorname{card} \mathscr{P}(X)$. The significance of this theorem is that it furnishes a way of constructing a far-reaching sequence of new (transfinite) cardinal numbers. For example, we have

$$\operatorname{card} \mathbf{R} < \operatorname{card} \mathscr{P}(\mathbf{R}) < \operatorname{card} \mathscr{P}(\mathscr{P}(\mathbf{R})) < \cdots.$$

Theorem 2. (*Cantor's Theorem*). If X is a set, then $\operatorname{card} X < \operatorname{card} \mathscr{P}(X)$.

Proof. If $X = \varnothing$, then $\operatorname{card} \varnothing = 0 < 1 = \operatorname{card} \mathscr{P}(\varnothing)$. Hence, it remains to prove the case where $X \neq \varnothing$. In this case, the function $g: X \to \mathscr{P}(X)$ given by $g(x) = \{x\} \in \mathscr{P}(X)$, for all $x \in X$, is injective. Thus, the set X is equipotent to the subset $\{\{x\} \mid x \in X\}$ of $\mathscr{P}(X)$ or, equivalently, $\operatorname{card} X \leq \operatorname{card} \mathscr{P}(X)$. From this, to show that $\operatorname{card} X < \operatorname{card} \mathscr{P}(X)$, it is sufficient to show that X is not equipotent to $\mathscr{P}(X)$.

Assume on the contrary that there is a bijection $f: X \sim \mathscr{P}(X)$; our aim is to prove that this assumption leads to a contradiction. Consider the set $S = \{x \in X \mid x \notin f(x)\}$. which consists of those elements of X that are not contained in their images under f. Since $S \in \mathscr{P}(X)$ and $f: X \sim \mathscr{P}(X)$, there exists an element $e \in X$ such that $f(e) = S$. Either $e \in S$ or $e \notin S$.

Case 1. $e \in S$.

It follows, by the definition of S, that $e \notin f(e)$; that is impossible, because $f(e) = S$ and $e \in S$.

Case 2. $e \notin S$.

Since $f(e) = S$, we have $e \notin f(e)$. Consequently, by the definition of S, $e \in S$ and hence $e \in f(e)$. This again is impossible.

A contradiction has been obtained and the proof of Cantor's Theorem is complete.

In view of Cantor's Theorem, a very natural question to arise was, is there a cardinal number x such that

$$\operatorname{card} \mathbf{N} < x < \operatorname{card} \mathscr{P}(\mathbf{N})$$

This question, called the *continuum problem*, captured the attention of Cantor and other mathematicians for a long time. More about this problem will be found in Section 8.

EXERCISE 6.3

1. Show that there is no largest cardinal number.
2. Let A and B be sets. Prove that if $A \sim B$ then $\operatorname{card} \mathscr{P}(A) = \operatorname{card} \mathscr{P}(B)$.

3. Let A be any denumerable set. Prove that the power set $\mathscr{P}(A)$ of A is nondenumerable.
4. Prove that the set of all infinite subsets of $\mathbf{N}$ is nondenumerable. [Hint: Use Problem 12, Exercise 5.3]
5. Prove by using Cantor's theorem (Theorem 2) that the set of all sets does not exist.

4. ADDITION OF CARDINAL NUMBERS

There is already an arithmetic for finite cardinal numbers. For instance, if k and l are two finite cardinal numbers, the sum $k + l$ and the product kl have their traditional meanings. We now try to generalize these concepts to cover the transfinite cardinal numbers as well; that is, to develop an arithmetic that applies to all cardinal numbers, finite or transfinite, and that will preserve the traditional meanings and properties of the arithmetic of finite cardinal numbers.

Definition 2. Let a and b be cardinal numbers. The *cardinal sum* of a and b, denoted by $a + b$, is the cardinal number $\operatorname{card}(A \cup B)$, where A and B are *disjoint* sets such that $\operatorname{card} A = a$ and $\operatorname{card} B = b$.

To show that Definition 2 is well-defined, the reader should first observe that for any two cardinal numbers a and b (not necessarily distinct), by the rule C-1 of Section 1, there exist sets X and Y such that $\operatorname{card} X = a$ and $\operatorname{card} Y = b$, where the sets X and Y need not be disjoint. But this causes no problem, since we may select $A = X \times \{0\}$ and $B = Y \times \{1\}$; then $A \sim X$, $B \sim Y$, and $A \cap B = \varnothing$. Thus, $a + b = \operatorname{card}(A \cup B)$ and this is uniquely defined; for if there are other disjoint set A' and B' such that $A' \sim A$ and $B' \sim B$, then by Theorem 6 of Chapter 5, we have $(A' \cup B') \sim (A \cup B)$ or, equivalently, $\operatorname{card}(A' \cup B') = \operatorname{card}(A \cup B)$.

We have thus proved the following theorem:

Theorem 3. Let a and b be cardinal numbers. Then

(a) There exist disjoint sets A and B such that $\operatorname{card} A = a$ and $\operatorname{card} B = b$.

(b) If A, B, A', and B' are sets such that $\operatorname{card} A' = \operatorname{card} A$, $\operatorname{card} B' = \operatorname{card} B$, $A \cap B = \varnothing$, and $A' \cap B' = \varnothing$, then $\operatorname{card}(A' \cup B') = \operatorname{card}(A \cup B)$.

The following example shows that Definition 2 agrees with the ordinary sum of two natural numbers when it is applied to two finite cardinal numbers.

EXAMPLE 2. Find the cardinal sum $4 + 3$ of the two finite cardinal numbers 4 and 3.

Solution. Since $\mathbf{N}_7 = \mathbf{N}_4 \cup \{5, 6, 7\}$, $\operatorname{card} \mathbf{N}_4 = 4$, $\operatorname{card}\{5, 6, 7\} = 3$, and the sets $\mathbf{N}_4$ and $\{5, 6, 7\}$ are disjoint, we have

$$\begin{aligned} 4 + 3 &= \operatorname{card}(\mathbf{N}_4 \cup \{5, 6, 7\}) \\ &= \operatorname{card} \mathbf{N}_7 = 7 \end{aligned}$$

which agrees with the ordinary sum of two integers.

Since the union of sets is commutative and associative we have the following corresponding properties about the cardinal sum.

Theorem 4. Let x, y, and z be arbitrary cardinal numbers. Then
(a) $x + y = y + x$ (Commutativity).
(b) $(x + y) + z = x + (y + z)$ (Associativity).

Following Georg Cantor, the symbols $\aleph_0$ (read *aleph-null*; $\aleph$ is the first letter of the Hebrew alphabet) and c have been used respectively to denote the cardinal number of a denumerable set and the cardinal number of the continuum, where continuum means the set of real numbers. In other words, $\aleph_0 = \operatorname{card} \mathbf{N}$ and $c = \operatorname{card} \mathbf{R}$.

EXAMPLE 3. Find the cardinal sum $\aleph_0 + \aleph_0$.

Solution. Let $\mathbf{N}_e$ and $\mathbf{N}_o$ denote, respectively, the set of even natural numbers and the set of odd natural numbers. Then, $\mathbf{N}_e$ and $\mathbf{N}_o$ are disjoint denumerable subsets of the set $\mathbf{N}$, and their union is $\mathbf{N}$. Consequently, by Definition 2,

$$\begin{aligned} \aleph_0 + \aleph_0 &= \operatorname{card} \mathbf{N}_e + \operatorname{card} \mathbf{N}_o \\ &= \operatorname{card}(\mathbf{N}_e \cup \mathbf{N}_o) \\ &= \operatorname{card} \mathbf{N} \\ &= \aleph_0 \end{aligned}$$

The result of Example 3 is a distinctive property of the transfinite cardinal numbers; for finite cardinal numbers, $n + m = n$ is true only for $m = 0$. The reader should prove, as an exercise, that $c + c = c$.

EXAMPLE 4. Find the cardinal sum $\aleph_0 + c$.

Solution. We have learned from Example 3, Section 2, Chapter 5, that the open interval (0, 1) and the set **R** of real numbers are equipotent. Hence, card(0, 1) = card **R** = c. Let **S** = **N** ∪ (0, 1). Then since **N** and (0, 1) are disjoint, card **S** = $\aleph_0 + c$. On the other hand, since **R** ~ (0, 1) ⊂ **S** and **S** ~ **S** ⊂ **R**, by the Schröder-Bernstein Theorem (Theorem 1), we have **S** ~ **R**. Therefore, $\aleph_0 + c = c$.

EXERCISE 6.4

1. Prove that $x + 0 = x$ for any cardinal number x.
2. Let x and y be two cardinal numbers. Prove that $x + y = y + x$.
3. Let x, y, and z be cardinal numbers. Prove that $(x + y) + z = x + (y + z)$.
4. Let n be an arbitrary finite cardinal number. Prove that
   (a) $n + \aleph_0 = \aleph_0$
   (b) $n + c = c$.
5. Prove that $c + c = c$.
6. Let x, y, and z be cardinal numbers.
   (a) Prove that if $x \leq y$ then $x + y \leq y + z$.
   (b) Show, by an example, that part (a) above is not true if "$\leq$" is replaced by "$<$."
7. Let x, y and z be cardinal numbers. Prove that if $x = y$ then $x + z = y + z$.
8. Show by giving a counterexample, that the converse of Problem 7, above, is not true.
9. Prove that card A + card B = card $(A \cup B)$ + card $(A \cap B)$ for any sets A and B.
10. Prove that if a is any transfinite cardinal number, then $a + 1 = a$. [Hint: Use the Schröder-Bernstein Theorem.]
11. Prove that if n is any finite cardinal number and a is any transfinite cardinal number, then $a + n = a$.
12. Prove the converse of Problem 11, above.

5. MULTIPLICATION OF CARDINAL NUMBERS

We now define the multiplication of cardinal numbers in such a way that for finite cardinal numbers, the result agrees with the ordinary multiplication of nonnegative integers.

Definition 3. For any cardinal numbers a and b, the *cardinal product ab* is defined to be the cardinal number of the Cartesian product $A \times B$, where card $A = a$ and card $B = b$.

To see that Definition 3 is independent of the choice of representatives A and B, let X and Y be sets such that $A \sim X$ and $B \sim Y$. Then, by Theorem 7 of Chapter 5, $A \times B \sim X \times Y$ and hence $\text{card}(A \times B) = \text{card}(X \times Y)$. It is also clear that this definition gives the right answer when a and b are finite cardinal numbers. Since we are all familiar with the multiplication of nonnegative integers, our main interest here is the product of transfinite cardinal numbers and the product of a finite and a transfinite cardinal number. First, let us list an easy consequence of Definition 3.

Theorem 5. Let x, y, and z be arbitrary cardinal numbers. Then

(a) $xy = yx$ (Commutativity).
(b) $(xy)z = x(yz)$ (Associativity).
(c) $x(y+z) = xy + xz$ (Distributivity).

Proof. Exercise.

EXAMPLE 5. Let x be an arbitrary cardinal number. Evaluate:

(a) $1x$
(b) $0x$
(c) $\aleph_0 \aleph_0$.

Solution. Let A be a set such that $\text{card} A = x$.

(a) Since the Cartesian product $\{1\} \times A$ is equipotent to the set A, we have $1x = x$.
(b) Since $\varnothing \times A = \varnothing$, we have $0x = 0$.
(c) Since $\mathbf{N} \times \mathbf{N} \sim \mathbf{N}$ (Theorem 10, Chapter 5), we have $\aleph_0 \aleph_0 = \aleph_0$.

EXAMPLE 6. Prove that $cc = c$, where $c = \text{card}\,\mathbf{R}$.

Proof. Since the set $\mathbf{R}$ and the open interval $(0, 1)$ of real numbers have the same cardinal number c, to show that $cc \leq c$, it is sufficient to show that there is an injection from the Cartesian product $(0, 1) \times (0, 1)$ to the interval $(0, 1)$. To this end, let us agree that each $x \in (0, 1)$ is expressed by its *infinite* decimal expansion so that, for example, the number ½ will be .4999 . . . but *not* .5. Thus we will have a unique expression for each number in $(0, 1)$. Now, we leave it to the reader to verify that the function $f\colon (0, 1) \times (0, 1) \to (0, 1)$ defined by

$$f(.x_1x_2x_3\cdots, .y_1y_2y_3\cdots) = .x_1y_1x_2y_2\cdots$$

is injective. This completes the proof that $cc \leq c$. The proof that $cc \geq c$ is left to the reader.

EXERCISE 6.5

1. Prove Theorem 5.
2. Let x, y, and z be cardinal numbers such that $x \leq y$. Prove that $xz \leq yz$.
3. Prove or disprove the following statement: If x, y, and z are cardinal numbers such that $x < y$ and $z \neq 0$, then $xz < yz$.
4. Let n be a finite cardinal number. Prove that $n\aleph_0 = \aleph_0$, if $n \neq 0$.
5. Let x and y be cardinal numbers. Prove that
   (a) If $xy = 0$ then $x = 0$ or $y = 0$.
   (b) If $xy = 1$ then $x = 1$ and $y = 1$.
6. Show that the function $f: (0, 1) \times (0, 1) \to (0, 1)$ defined by
$$f(.x_1x_2x_3\cdots, .y_1y_2y_3\cdots) = .x_1y_1x_2y_2\cdots$$
   in the proof of Example 6 is not surjective
7. Prove that if x is any cardinal number and n is a finite cardinal number, then $nx = x + x + x + \cdots + x$ (n summand).
8. Let x, y, and z be any cardinal numbers. Prove that if $x = y$ then $xz = yz$.
9. Show by giving a counterexample that the converse of Problem 8, above, is not true.

6. EXPONENTIATION OF CARDINAL NUMBERS

Let a and b be cardinal numbers, finite or transfinite. In order to give a satisfactory meaning to b^a (*read:* ath power of b), we first examine the finite case: $2^3 = 2\cdot 2\cdot 2$ and, in general, $n^m = n\cdot n\cdot\cdots\cdot n$ (m factor). We could generalize this concept to the transfinite case by introducing "generalized Cartesian products," but there is an approach that works without reference to generalized Cartesian products. Let A be a set with m elements and B a set with n elements. How many functions are there from A to B (see Problem 11, Exercise 3.4)? Since each element of A has n choices for its image, and this selection of image may be made independently m times (once for each element of A), the answer is $n\cdot n\cdot\cdots\cdot n = n^m$. This concept is generalized as follows:

Definition 4. Let a and b be cardinal numbers with $a \neq 0$. Let A and B be sets such that $\operatorname{card} A = a$ and $\operatorname{card} B = b$. Denote the set of all functions from A to B by B^A. We define $b^a = \operatorname{card} B^A$.

Before we can accept Definition 4, we need to verify that this definition is independent of the choice of representatives A and B. The following theorem is what is needed.

Theorem 6. Let A, B, X, and Y be sets such that $A \sim X$, $B \sim Y$. Then $B^A \sim Y^X$.

Proof. Let $g : A \sim X$ and $h : B \sim Y$ be bijections. Then we define the function

$$\psi : B^A \to Y^X$$

by $\psi(f) : X \to Y$, where $\psi(f)(x) = h \circ f \circ g^{-1}(x)$ for all $f \in B^A$.

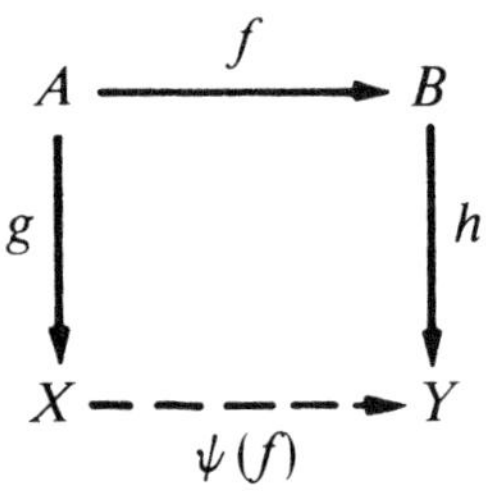

We leave it to the reader to prove that the function $\psi : B^A \to Y^X$ is bijective.

EXAMPLE 7. Let A be a set. Compare the cardinal numbers card $\mathscr{P}(A)$ and $2^{\text{card} A}$.

Solution. Let $B = \{0,1\}$. We assign to each subset D of A the characteristic function $\chi_D : A \to B$ defined in Example 8, Chapter 3. The function from $\mathscr{P}(A)$ to B^A which sends D to χ_D is bijective (Prove it!). Thus, the sets $\mathscr{P}(A)$ and B^A have the same cardinal number; that is, card $\mathscr{P}(A) = \text{card}\, B^A = 2^{\text{card} A}$.

Theorem 7. Let a, x, and y be cardinal numbers. Then $a^x a^y = a^{x+y}$.

Proof. Let A, X, and Y be sets such that $\text{card}\, A = a$, $\text{card}\, X = x$, $\text{card}\, Y = y$, and $X \cap Y = \varnothing$. Then, by Definition 2, $\text{card}(X \cup Y) = x + y$. It is sufficient to show that the sets $A^X \times A^Y$ and $A^{X \cup Y}$ are equipotent. To this end, we assign to each pair (f, g) of functions, $f \in A^X$ and $g \in A^Y$, the function $f \cup g \in A^{X \cup Y}$ [see Theorem 8, Chapter 3]. We leave it to the reader to verify that this assignement establishes an equipotence between the sets $A^X \times A^Y$ and $A^{X \cup Y}$. Hence, $a^x a^y = a^{x+y}$.

Theorem 8. Let x, y, and z be cardinal numbers. Then $(z^y)^x = z^{yx}$.

Proof. Let X, Y, and Z be sets with cardinal numbers x, y, and z, respectively According to Definition 4, the theorem is proved if we establish that $Z^{Y \times X} \sim (Z^Y)^X$. Before showing this equipotence, we first need a notational convention: For a given function $f: Y \times X \to Z$ and a given element $a \in X$, there exists a function $f^a : Y \to Z$ defined by $f^a(b) = f(b, a)$ for all $b \in Y$. We leave it to the reader to show that the function $\psi : Z^{Y \times X} \to (Z^Y)^X$ which assigns to each $f \in Z^{Y \times X}$ the function $e_f \in (Z^Y)^X$ given by $e_f(a) = f^a$ for all $a \in X$ is a bijection.

Recall that the A-projection $p_A : A \times B \to A$ is a function that assigns a to each ordered pair $(a, b) \in A \times B$; the B-projection $p_B : A \times B \to B$ is similarly defined [see Problem 8, Exercise 3.5]

Theorem 9. Let a, b, and x be cardinal numbers. Then $(ab)^x = a^x b^x$.

Proof. Let A, B, and X be set with cardinal numbers a, b, and x, respectively. The function $\psi : (A \times B)^X \to A^X \times B^X$ which pairs each $f: X \to A \times B$ with the function $(p_A \circ f, p_A \circ f)$ in $A^X \times B^X$ is bijective (Prove it!). Hence, by Definition 4, $(ab)^x = a^x b^x$.

Recall that the symbols $\aleph_0$ and c denote the cardinal numbers of the sets $\mathbf{N}$ and $\mathbf{R}$, respectively, and that $\mathbf{Q} \sim \mathbf{N}$ (see Example 5, Chapter 5) and $(0, 1) \sim \mathbf{R}$ (see Example 3, Chapter 5). Thus $\aleph_0$ is the cardinal number of $\mathbf{Q}$ and c is the cardinal number of the interval $(0, 1)$.

Theorem 10. $2^{\aleph_0} = c$.

Proof. We shall prove this in two steps, showing first that $c \leq 2^{\aleph_0}$ and then that $2^{\aleph_0} \leq c$.

Consider the function $f: \mathbf{R} \to \mathscr{P}(\mathbf{Q})$ defined by

$$f(a) = \{x \in \mathbf{Q} \mid x < a\}, \qquad \text{for all} \quad a \in \mathbf{R}$$

This function is injective: If $a < b$ are two distinct real numbers, then there exists a rational number r such that $a < r < b$.[3] Then $r \in f(b)$ but $r \notin f(a)$, and hence f is injective. This proves, by using the results of Problem 3, Exercise 6.2, and Example 7, that

$$c \leq \operatorname{card} \mathscr{P}(\mathbf{Q}) = 2^{\aleph_0}$$

To prove the reverse inequality, let $\psi : \{0, 1\}^{\mathbf{N}} \to \mathbf{R}$ be the function given by

[3] Because the set of rational numbers is a dense subset of the real numbers.

$$\psi(f) = 0.f(1)f(2)f(3)\cdots$$

where $f \in \{0, 1\}^{\mathbf{N}}$. Note that $\psi(f)$ is a decimal number (consisting of 0's and 1's). If $f, g \in \{0, 1\}^{\mathbf{N}}$ and $f \neq g$, then $\psi(f) \neq \psi(g)$ because the decimals which define $\psi(f)$ and $\psi(g)$ are distinct. Therefore, $\psi : \{0, 1\}^{\mathbf{N}} \to \mathbf{R}$ is injective, and hence $2^{\aleph_0} \leq c$.

Corollary. $\aleph_0 < c$.

Proof. By Cantor's Theorem (Theorem 2) and by the result of Example 7, we have

$$\aleph_0 < \text{card}\,\mathcal{P}(\mathbf{N}) = 2^{\text{card}\,\mathbf{N}} = 2^{\aleph_0} = c$$

EXERCISE 6.6

1. Prove that the function $\psi : B^A \to Y^X$ in the proof of Theorem 6 is bijective.
2. Let a be an arbitrary cardinal number. Prove that $a^0 = 1$, $a^1 = a$, $1^a = 1$, and $0^a = 0$ if $a \neq 0$.
3. Show that $2^a > a$ for any cardinal number a.
4. Let a, b, x, and y be cardinal numbers such that $a \leq b$ and $0 < x \leq y$. Prove that $a^x \leq b^y$.
5. Prove that $n^{\aleph_0} = c = \aleph_0^{\aleph_0}$ for any finite $n \geq 2$.
6. Prove that $c^{\aleph_0} = c = c^n$ for any finite $n \geq 1$.
7. Let C denote the set of all complex numbers. Prove that $\text{card}\,C = c$.
8. Prove that $\aleph_0 c = c$.
9. Prove that the function from $\mathcal{P}(A)$ to $\{0, 1\}^A$ which pairs each D in $\mathcal{P}(A)$ with χ_D is bijective.
10. Let A, X, and Y be sets such that X and Y are disjoint. Prove that the function from $A^X \times A^Y$ to $A^{X \cup Y}$ which pairs each (f, g) in $A^X \times A^Y$ with $f \cup g$ in $A^{X \cup Y}$ is bijective.
11. Prove that the function $\psi : Z^{Y \times X} \to (Z^Y)^X$ in the proof of Theorem 8 is bijective.
12. Prove that the function $\psi : (A \times B)^X \to A^X \times B^X$ in the proof of Theorem 9 is bijective.

7. FURTHER EXAMPLES OF CARDINAL ARITHMETIC

In Example 6, it was proved directly that $cc = c$. Now, using Theorem 10, $2^{\aleph_0} = c$, a shorter proof can be given.

EXAMPLE 8. Prove that $cc = c$ by using Theorem 10 [cf. Example 6].

Proof. It follows from Theorems 7 and 10 and Example 3, $\aleph_0 + \aleph_0 = \aleph_0$, that

$$cc = 2^{\aleph_0} 2^{\aleph_0} = 2^{\aleph_0 + \aleph_0} = 2^{\aleph_0} = c$$

EXAMPLE 9. Compare the cardinal number of the set $\{f \mid f: \mathbf{R} \to \mathbf{R}\}$ of all functions from $\mathbf{R}$ to $\mathbf{R}$ and the cardinal number c of $\mathbf{R}$.

Solution. We have

$\operatorname{card}\{f \mid f: \mathbf{R} \to \mathbf{R}\}$	$= c^c$	Def. 4
	$= (2^{\aleph_0})^c$	Th. 10
	$= 2^{\aleph_0 c}$	Th. 8
	$= 2^c$	Prob. 8, Ex. 6.6
	$> c$	Example 7, Th.2

Therefore, $\operatorname{card}\{f \mid f: \mathbf{R} \to \mathbf{R}\} > \operatorname{card} \mathbf{R}$.

EXAMPLE 10. Let $C(\mathbf{R}, \mathbf{R})$ and $C(\mathbf{Q}, \mathbf{R})$ be the sets of continuous real-valued functions with domain $\mathbf{R}$ and domain $\mathbf{Q}$, respectively. Let $K(\mathbf{R}, \mathbf{R})$ be the set of all real-valued constant functions with domain $\mathbf{R}$. Prove that

$$\operatorname{card} C(\mathbf{R}, \mathbf{R}) = \operatorname{card} C(\mathbf{Q}, \mathbf{R}) = \operatorname{card} K(\mathbf{R}, \mathbf{R}) = c$$

Proof.[4] To each function $f: \mathbf{R} \to \mathbf{R}$, there corresponds a function $f \mid \mathbf{Q}: \mathbf{Q} \to \mathbf{R}$ defined by $(f \mid \mathbf{Q})(x) = f(x)$ for all $x \in \mathbf{Q}$. The function $f \mid \mathbf{Q}$ is called the *restriction* of f to $\mathbf{Q}$. Hence, there is a natural function

$$\psi: C(\mathbf{R}, \mathbf{R}) \to C(\mathbf{Q}, \mathbf{R})$$

which takes each $f \in C(\mathbf{R}, \mathbf{R})$ to its restriction $f \mid \mathbf{Q}$ on $\mathbf{Q}$. It is clear that the restriction of a continuous function is continuous. Therefore, $\psi: C(\mathbf{R}, \mathbf{R}) \to C(\mathbf{Q}, \mathbf{R})$ is a well-defined function.

It follows from the density property of the rational numbers in the real numbers that for each real number x there is a sequence $\{x_n \mid n \in \mathbf{N}\}$ of rational numbers such that

$$\lim_{n \to \infty} x_n = x$$

Consequently, if any two continuous functions $f, g: \mathbf{R} \to \mathbf{R}$ have the property that $f(x') = g(x')$ for all $x' \in \mathbf{Q}$, then $f(x) = g(x)$ for all $x \in \mathbf{R}$. In other words, the function $\psi: C(\mathbf{R}, \mathbf{R}) \to C(\mathbf{Q}, \mathbf{R})$ is injective. Therefore, we have

[4] The proof may be omitted at the discretion of the instructor.

$$\begin{aligned}\operatorname{card} C(\mathbf{R}, \mathbf{R}) &\leq \operatorname{card} C(\mathbf{Q}, \mathbf{R}) \\ &\leq \operatorname{card} \mathbf{R}^{\mathbf{Q}} \\ &= c^{\aleph_0} \\ &= (2^{\aleph_0})^{\aleph_0} \\ &= 2^{(\aleph_0 \aleph_0)} \\ &= 2^{\aleph_0} \\ &= c\end{aligned}$$

by Theorems 8 and 10.

Now, consider the set $K(\mathbf{R}, \mathbf{R})$ of all real-valued constant functions with the domain $\mathbf{R}$. Since to each real number a, there exists a constant function $f_a : \mathbf{R} \to \mathbf{R}$ defined by $f_a(\mathbf{R}) = \{a\}$, we have

$$\operatorname{card} K(\mathbf{R}, \mathbf{R}) = c$$

Since each constant function $f_a : \mathbf{R} \to \mathbf{R}$ is continuous, we have $K(\mathbf{R}, \mathbf{R}) \subset C(\mathbf{R}, \mathbf{R})$. Therefore,

$$c = \operatorname{card} K(\mathbf{R}, \mathbf{R}) \leq \operatorname{card} C(\mathbf{R}, \mathbf{R})$$

which combined with the inequality obtained in the last paragraph, gives

$$c = \operatorname{card} K(\mathbf{R}, \mathbf{R}) \leq \operatorname{card} C(\mathbf{R}, \mathbf{R}) \leq \operatorname{card} C(\mathbf{Q}, \mathbf{R}) \leq c$$

This completes the proof that $\operatorname{card} C(\mathbf{R}, \mathbf{R})$, $\operatorname{card} C(\mathbf{Q}, \mathbf{R})$, and $\operatorname{card} K(\mathbf{R}, \mathbf{R})$ are equal to c.

The result of Example 10 indicates that the constant functions are as "numerous" as the continuous functions. This is another illustration of the curious properties of infinite sets.

EXAMPLE 11. Find the cardinal number of the set $D(\mathbf{R}, \mathbf{R})$ of all differentiable real-valued functions of a real variable.

Solution. Since each constant function is differentiable and each differentiable function is continuous, we have

$$K(\mathbf{R}, \mathbf{R}) \subseteq D(\mathbf{R}, \mathbf{R}) \subseteq C(\mathbf{R}, \mathbf{R})$$

By Example 10, we have

$$c = \operatorname{card} K(\mathbf{R}, \mathbf{R}) \leq \operatorname{card} D(\mathbf{R}, \mathbf{R}) \leq \operatorname{card} C(\mathbf{R}, \mathbf{R}) = c$$

Therefore, $\operatorname{card} D(\mathbf{R}, \mathbf{R}) = c$.

EXERCISE 6.7

1. Show that the n-dimensional space $\mathbf{R}^n = \mathbf{R} \times \mathbf{R} \times \cdots \times \mathbf{R}$ (n factors)

contains "just as many" points as the unit open interval (0, 1).

2. The classic *Hilbert space* consists of all infinite sequences $(x_1,x_2,x_3,\ldots)$ of real numbers, called *points*, for which the series $x_1^2+x_2^2+x_3^2+\cdots$ converges. Show that the classic Hilbert space contains "just as many" points as the real line **R**.
3. Let $\mathbf{R}^{\aleph_0}$ be the set of all infinite sequences $(x_1, x_2, x_3, \ldots)$ of real numbers, call *points* of the space $\mathbf{R}^{\aleph_0}$. A *lattice point* in $\mathbf{R}^{\aleph_0}$ is a point $(x_1, x_2, x_3, \ldots)$ such that all x_k's are integers. Show that the space $\mathbf{R}^{\aleph_0}$ contains "just as many" points as the set of all lattice points in $\mathbf{R}^{\aleph_0}$.
4. Show that there are "just as many" functions of one real variable which assume only the values 0 and 1 as all real-valued functions of n real variables, where n is any natural number.
5. Let the cardinal number of the set $\{f \mid f: \mathbf{R} \to \mathbf{R}\}$ of all real-valued functions of one variable be denoted by $\mathfrak{f}$. Show that

$$\mathfrak{f}^n = \mathfrak{f}^{\aleph_0} = \mathfrak{f}^c = \mathfrak{f}$$

for all $n \in \mathbf{N}$.

8. THE CONTINUUM HYPOTHESIS AND ITS GENERALIZATION

Since every infinite set contains a denumerable subset (Theorem 11, Chapter 5), the cardinal number $\aleph_0$ is the smallest transfinite cardinal number. An important question, known as the *continuum problem*, was raised by Cantor about 1880: Is there a cardinal number that lies strictly between $\aleph_0$ and $2^{\aleph_0}$ $(=c)$? In set language, are there any nondenumerable infinite subsets of **R** with cardinal number less than that of **R**? Cantor and many leading mathematicians had tried in vain to solve this problem. Since no such set had been found anywhere in classical mathematics and there seemed to be no way of finding one, it was conjectured by Cantor and others that the answer to the continuum problem must be no. This conjecture is known as the *continuum hypothesis*.

Continuum Hypothesis. There is no cardinal number x satisfying $\aleph_0 < x < c$ $(= 2^{\aleph_0})$.

A question closely related to the continuum problem, usually referred to as the *generalized continuum problem*, is the following: Is there any cardinal number that lies strictly between a transfinite cardinal number a and 2^a? This question too has not been answered. The conjecture that no such cardinal number exists is called the *generalized continuum hypothesis*.

Generalized Continuum Hypothesis. For any transfinite cardinal number a, there is no cardinal number x such that $a < x < 2^a$.

As early as 1900, at the International Congress of Mathematicians in Paris, the great German mathematician David Hilbert[5] (1862-1943) presented a list of 23 important unsolved mathematical problems, the first of which was the continuum problem. There was no progress made in solving this problem until 1938, when Kurt Gödel[6] (1906-1978), the outstanding logician of the century, proved that if the generalized continuum hypothesis is added to the current axioms for set theory, then any contradiction that might be implied by this system of axioms could be formulated as a contradiction implied by the initial axioms (without the generalized continuum hypothesis) alone.[7] In other words, the generalized continuum hypothesis is relatively consistent with the axioms of set theory.

Finally, in 1963, a significant achievement was made by the young mathematician Paul J. Cohen (1934-) of Stanford University, who declared that the generalized continuum hypothesis is unprovable on the basis of the current axioms for set theory. Therefore, the status of the generalized continuum hypothesis in set theory is analogous to that of Euclid's parallel axiom (the Fifth Postulate) in geometry. We may postulate them or deny them, in either case obtaining a consistent theory of mathematics.

[5] David Hilbert (1862-1943), an outstanding mathematician of all time, was a professor of mathematics at the University of Göttingen Germany (1895-1943). He influenced the entire world of modern mathematics, ranging from 19th century algebra to modern logic and mathematical physics. The famous Hilbert space is one of his many contributions. Hilbert believed that all mathematical ideas fit together harmoniously.

[6] Kurt Gödel (1906-1978) of the Princeton Institute for Advanced Study in New Jersey was born in Czechoslovakia. He first achieved fame at 25. Famous scholars including Bertrand Russell (1872-1970) and Alfred North Whitehead (1861-1947) had suggested the existence of absolute guides to the truth or falsity of certain mathematical statements. Gödel shocked the world by proving that what Russell and Whitehead sought did not exist. His other major contributions include the proof of the completeness of quantification logic and the proof of the consistency of both the generalized continuum hypothesis and the axiom of choice.

[7] See K. Gödel, *The Consistency of the Axiom of Choice and of the Generalized Continuum Hypothesis with the Axioms of Set Theory*, Princeton University Press, Princeton, N.J., 1940, 66 pp. Rev. ed., 1951, 74 pp.

Chapter Seven

The Axiom of Choice and Some of its Equivalent Forms

The axiom of choice and three frequently used equivalent principles — Hausdorff maximality principle, Zorn's lemma, and the well-ordering principle of Zermelo — are introduced. It is proved, in a circular chain of implications, that the axiom of choice and these three important mathematical principles are all logically equivalent. The principle of transfinite induction is given together with a demonstration of how this induction may be used in mathematical proofs. A short historical comment of the axiom of choice ends the chapter.

1. INTRODUCTION

Suppose that you enter a fruit market that has a number of (nonempty) baskets of fruit. If you are allowed to choose one (and only one) fruit from each basket as free samples, you know this would not be a difficult task. But the following similar question, which may be seemingly trivial at first glance, is really complex:

Given a nonempty set $\mathcal{S}$ whose elements are disjoint nonempty sets S_α, does there exist a set R which has as its elements one element x_α of each S_α?

The real difficulty lies in the case where $\mathcal{S}$ is infinite. At the beginning of the twentieth century, Ernst Zermelo (1871—1953) and others made unsuccessful attempts to answer this question. Zermelo felt that this question might be unsolvable and that the only way to avoid the difficulty was to postulate an axiom, which has since been known as the *axiom of choice*. We state this axiom in its general form in terms of functions:

Axiom of Choice. To any nonempty set $\mathcal{S}$ whose elements are nonempty sets, there exists a function called a *choice function* $f\colon \mathcal{S} \to \bigcup_{A \in \mathcal{S}} A$ such that $f(A) \in A$ for all $A \in \mathcal{S}$.

The axiom of choice is now indispensable for proving many important results in various areas of contemporary mathematics. In fact, a disguised use of the axiom of choice was made earlier in the proof of Theorem 11, Chapter 5.

Mathematicians have since discovered several other principles that often can be used as convenient substitutes for the axiom of choice. These principles, though seemingly bearing no resemblance to the axiom of choice, were soon shown to be equivalent to the axiom of choice. To understand some of these principles and their connection with the axiom of choice, we need a few definitions.

Definition 1. A relation $\leq$ on a set A is called a *partial order* relation if and only if the relation $\leq$ is reflexive and transitive on A and *antisymmetric* on A: that is, if $a \leq b$ and $b \leq a$ then $a = b$. A *partially ordered set* is a pair $(A, \leq)$, where A is a set and $\leq$ is a partial order relation on A.

Definition 2. A *total order* relation $\leq$ on a set A is a partial order relation such that for any pair of elements a and b in A, either $a \leq b$ or $b \leq a$. A *totally* ordered set is a pair $(A, \leq)$ where A is a set and $\leq$ is a total order relation.

When the partial (total) order relation $\leq$ on A is unmistakably clear from the context, we may simply say that A is a partially (totally) ordered set. Total order relations and totally ordered set are also called *linear order relations* and *linearly ordered sets*, respectively. It is evident from the definitions that a totally ordered set is a partially ordered set, but a partially ordered set need not be a totally ordered set (see Example 2, below). Let B be a subset of a partially ordered set $(A, \mathcal{R})$, and let $\leq_B$ be the intersection $\mathcal{R} \cap (B \times B)$ of $\mathcal{R}$ with $B \times B$. Then $(B, \leq_B)$ is a partially ordered set; it may happen that $\leq_B$ is a total order relation on B, in which case B is called a *totally ordered subset* of the partially ordered set $(A, \mathcal{R})$. A totally ordered subset of a partially ordered set is called a *chain*.

If $(A, \leq)$ is a partially ordered (totally ordered) set, we may say, equivalently, that the set A is partially ordered (totally ordered) by $\leq$. In either case, we may write $a \geq b$ for $b \leq a$, and $a < b$ or $b > a$ for $a \leq b$ and $a \neq b$.

EXAMPLE 1. Let X be a nonempty set. The power set $\mathcal{P}(X)$ of X is partially ordered by the inclusion relation $\subseteq$ on $\mathcal{P}(X)$.

EXAMPLE 2. Let $\leq'$ be the relation defined on $\mathbf{R}^2$ by $(a_1, a_2) \leq' (b_1, b_2)$ if and only if $a_1 \leq b_1$ and $a_1 \leq b_2$. The reader should verify that the relation $\leq'$ is a partial order relation on $\mathbf{R}^2$. Since neither $(1, 2) \leq' (2, 1)$ nor $(2, 1) \leq' (1, 2)$, the relation $\leq'$ is not a total order relation on $\mathbf{R}^2$.

EXAMPLE 3. In Example 2, the diagonal $\Delta = \{(x, x) \mid x \in \mathbf{R}\}$ of the plane $\mathbf{R}^2$ is a chain.

EXERCISE 7.1

1. Let $\mathscr{F}$ be a partition of a nonempty set X. Prove that there exists a set $R \subseteq X$ such that for every $C \in \mathscr{F}$, $C \cap R$ consists of one and only one element; such a set R is called a set of *representatives* for $\mathscr{F}$.
2. Let $f: A \to B$ be a surjection. Prove that there exists a subset C of A such that C is equipotent to B and hence $\operatorname{card} A \geq \operatorname{card} B$.
3. Prove the statement of Example 1.
4. Show that
   (a) the identity relation "$=$" on a set is a partial order relation.
   (b) the ordinary "$\leq$" relation on the set of real numbers is a total order relation.
5. Let F be the set of all functions $f: \mathbf{R} \to \mathbf{R}$, and let
$$\mathscr{R} = \{(f, g) \in F \times F \mid f(x) \leq g(x), \ \forall\, x \in \mathbf{R}\}$$
   Prove that $(F, \mathscr{R})$ is a partially ordered set.
6. Prove that the set $\mathbf{N}$ of natural numbers together with the relation "x divides y" is a partially ordered set.
7. Let $(\mathbf{R}^2, \leq')$ be the partially ordered set of Example 2, and let m be any nonnegative real number. Prove that the subset $\{(x, mx) \mid x \in \mathbf{R}\}$ of $(\mathbf{R}^2, \leq')$ is a chain.
8. Let $\{A_\gamma \mid \gamma \in \Gamma\}$ be a family of nonempty sets; that is, each $A_\gamma \neq \varnothing$. Then the *generalized Cartesian product* $\mathbf{P}_{\gamma \in \Gamma} A_\gamma$ of the family $\{A_\gamma \mid \gamma \in \Gamma\}$ is defined to be the set of all functions $f: \Gamma \to \bigcup_{\gamma \in \Gamma} A_\gamma$ such that $f(\gamma) \in A_\gamma$ for all $\gamma \in \Gamma$. Prove that if $\Gamma \neq \varnothing$ then $\mathbf{P}_{\gamma \in \Gamma} A_\gamma$ is not empty.
9. Using the same notation as Problem 8, prove that if $\Gamma = \{1, 2, \ldots, n\}$ is finite then there is a one-to-one correspondence between $\mathbf{P}_{\gamma \in \Gamma} A_\gamma$ and the set $\{(x_1, x_2, \ldots, x_n) \mid x_i \in A_i \text{ for all } i \in \Gamma\}$ of n-tuples.
10. Let A and B be nonempty sets and let $\mathscr{R}$ be a relation from A to B *with Dom* $\mathscr{R} = A$. Prove that there exists a function $f: A \to B$ such that $f \subseteq \mathscr{R}$.
11. Prove that a function $f: A \to B$ is surjective if and only if there exists a function $g: B \to A$ such that $f \circ g = I_B$ the identity function on B [cf. Theorem 16, Chapter 3].

2. THE HAUSDORFF MAXIMALITY PRINCIPLE

In modern algebra and in topology, we often find it more convenient to use the equivalent Hausdorff maximality principle than to use the axiom of choice. In order to understand this principle, a few more new terms are needed.

Definition 3. Let $(A, \leq)$ be a partially ordered set.

(a) An element $u \in A$ is said to be an *upper bound* for a subset B of A if and only if $u \geq b$ for all $b \in B$.

(b) An upper bound u_0 for B is the *least upper bound* for B if and only if $u_0 \leq u$ for every upper bound u for B.

(c) An element $e \in A$ is said to be *maximal* if and only if $e \leq a$ implies $e = a$ for all $a \in A$.

Definition 3 has a dual form which we state separately as

Definition 3′. Let $(A, \leq)$ be a partially ordered set.

(a) An element $v \in A$ is a *lower bound* for a subset B of A if and only if $v \leq b$ for all $b \in B$.

(b) A lower bound v_0 for B is the *greatest lower bound* for B if and only if $v_0 \geq v$ for every lower bound v for B.

(c) An element $e' \in A$ is *minimal* if and only if $a \leq e'$ implies $e' = a$ for all $a \in \mathbf{A}$.

EXAMPLE 4. Let X be a nonempty set and let $\mathcal{B}$ be a subset of the partially ordered set $(\mathcal{P}(X), \subseteq)$ [see Example 1]. Then the least upper bound of $\mathcal{B}$ is $\bigcup_{B \in \mathcal{B}} B$, and the greatest lower bound of $\mathcal{B}$ is $\bigcap_{B \in \mathcal{B}} B$.

The following theorem plays an important role in proving the equivalence of the axiom of choice to other principles. The proof of this theorem is tedious, and perhaps discouraging; therefore the authors suggest that beginners take this theorem for granted, and omit the proof in their first reading.

Theorem 1. Let $(A, \leq)$ be a nonempty partially ordered set such that every totally ordered subset of A has a least upper bound in A. If $f: A \to A$ is such that $f(a) \geq a$ for every $a \in A$, then there exists $p \in A$ such that $f(p) = p$.

Proof. Let a be an arbitrary element of A that remains fixed throughout the proof. A subset B of A is called *admissible* if and only if it has the following three properties:

(i) $a \in B$.

(ii) $f(B) \subseteq B$.

(iii) The least upper bound of each nonempty totally ordered subset of B belongs to B.

Let $\mathcal{B}$ be the set of all admissible subsets of A. Then, since the set A itself is admissible, $\mathcal{B} \neq \emptyset$. An intersection of admissible sets is admissible; hence the partially ordered set $(\mathcal{B}, \subseteq)$ has a unique minimal element $B_0 = \bigcap_{B \in \mathcal{B}} B$. Since the set $\{x \in A \mid x \geq a\}$ is clearly admissible, we have $B_0 \subseteq \{x \in A \mid x \geq a\}$. Thus

(iv) $x \geq a$ for all $x \in B_0$.

Let $C = \{x \in B_0 \mid y \in B_0 \text{ and } y < x \text{ imply } f(y) \leq x\}$. We shall prove that

(v) $x \in C$ and $z \in B_0$ imply either $z \leq x$ or $z \geq f(x)$.

Fix $x \in C$, and denote $D = \{z \in B_0 \mid z \leq x \text{ or } z \geq f(x)\}$. Then the condition (iv) shows that D satisfies (i). The set D satisfies (ii): for if $z \geq f(x)$, then $f(z) \geq z \geq f(x)$; if $z = x$, then $f(z) = f(x)$; and if $z < x$, then since $x \in C$ we have $f(z) \leq x$. The set D also satisfies (iii): If u is a least upper bound for the nonempty totally ordered subset E of D, then either $y \leq x$ for all $y \in E$, and consequently $u \leq x$, or $y \geq f(x)$ for some $y \in E$, and consequently $u \geq f(x)$. Thus, D is admissible and hence $D = B_0$, which proves the statement (v).

We now prove that C is admissible. The set C satisfies (i) vacuously. To show that C satisfies (ii), we show that if $x \in C$, $y \in B_0$, and $y < f(x)$, then $f(y) \leq f(x)$. By (v), we have either $y \geq f(x)$ or $y \leq x$, so that if $y < f(x)$ then $y \leq x$. Since $x \in C$, $y < x$ implies $f(y) \leq x$ $(\leq f(x))$. If $y = x$ then $f(y) = f(x)$. To prove that C satisfies (iii), let w be the least upper bound for the nonemtpy totally ordered set $G \subseteq C$. To verify that $w \in C$, let $y \in B_0$ and $y < w$. We have, by (v), that each $x \in G$ has the property that either $y \leq x$ or $y \geq f(x)$. The inequality $y \geq f(x) \geq x$ cannot be true for all $x \in G$, for then $y \geq w$, which contradicts the choice of y. Therefore, there exists an $x \in G$ such that $y \leq x$. If $y < x$ then, by the definition of C, $f(y) \leq x \leq w$. If $y = x$ then, since $y < w$, there exists a $z \in G$ such that $y < z$. Whence, by definition of $C, f(y) \leq z \leq w$. Thus, $f(y) \leq w$ and hence $w \in C$. Now, since the set C is an admissible subset of B_0, we must have $C = B_0$. Consequently, by (v), B_0 is totally ordered. Let p be the least upper bound of B_0; then $p \in B_0$ and $p \leq f(p) \leq p$, so that $f(p) = p$.

We are now ready to prove the Hausdorff Maximality Principle by use of the axiom of choice.

Theorem 2. *(Hausdorff Maximality Principle).* Let the set $\mathcal{T}$ of all totally ordered subsets of a partially ordered set $(A, \leq)$ be partially ordered by inclusion, $\subseteq$. Then $(\mathcal{T}, \subseteq)$ has a maximal element.

Proof. Suppose on the contrary that $\mathcal{T}$ has no maximal element. Then to each $T \in \mathcal{T}$, there is associated a nonempty set

$$T^* = \{S \in \mathcal{T} \mid S \supset T\}$$

By the axiom of choice, there is a function g with domain $\{T^* \mid T \in \mathcal{T}\}$

satisfying $g(T^*) \in T^*$. Consequently, there is a function $f\colon \mathscr{T} \to \mathscr{T}$ defined by $f(T) = g(T^*) \supset T$ for all $T \in \mathscr{T}$. Then using Example 4 we see that $(\mathscr{T}, \subseteq)$ together with the function f satisfies the hypotheses of Theorem 1. But $f(T) \supset T$ for all $T \in \mathscr{T}$, a contradiction. Thus, the theorem is proved.

EXERCISE 7.2

1. Let B be a subset of the partially ordered set $(A, \leq)$. Prove that a least upper bound (greatest lower bound) of B is unique if it exists.
2. Give an example of a bounded subset of a partially ordered set that has neither a least upper bound nor a greatest lower bound.
3. Let $X = \{a, b, c\}$ and let the power set $\mathscr{P}(X)$ be partially ordered by the inclusion $\subseteq$. Find all upper bounds, all lower bounds, the least upper bound, and the greatest lower bound for the set $\{\{a, b\}, \{c, a\}\}$.
4. Prove the statement of Example 4.
5. In Problem 3 find maximal and minimal elements of $\mathscr{P}(X)$.
6. Find maximal and minimal elements of the partially ordered set $\mathscr{P}(X)$ in Example 4.
7. Give an example of a partially ordered set that has more than one maximal element and more than one minimal element.
8. Prove that if a totally ordered set has a maximal (minimal) element, then it has a unique maximal (minimal) element.
9. Give as example of a totally ordered set that has neither a maximal element nor a minimal element.
10. Let $(\mathscr{T}, \subseteq)$ be as in Theorem 2, and let $\mathscr{T}_0$ be a totally ordered subset of $\mathscr{T}$. Prove that $T_0 = \bigcap\{T \in \mathscr{T} \mid T \supseteq S \text{ for all } S \in \mathscr{T}_0\}$ is the least upper bound of $\mathscr{T}_0$.
11. Prove the following. Let $(A, \leq)$ be a nonempty partially ordered set such that every totally ordered subset of A has a greatest lower bound. If $f\colon A \to A$ is such that $f(a) \leq a$, for every $a \in A$, then there exists $q \in A$ such that $f(q) = q$.
[Hint: Use Theorem 1.]
12. Let $(A, \leq)$ be a partially ordered set and let B be a totally ordered subset of A. Prove that A has a maximal totally ordered subset C such that $B \subseteq C$. [Hint: Use Theorem 2.]
13. Let $(A, \leq)$ be a partially ordered set. A subset B of A is called an *antichain* if and only if for any two distinct elements x and y in B, neither $x \leq y$ nor $y \leq x$. Prove that every antichain is contained in an antichain which is maximal with respect to inclusion $\subseteq$. [Hint: Use Theorem 2.]

3. ZORN'S LEMMA

Prehaps one of the most widely used equivalent forms of the axiom of choice is Zorn's lemma, which first appeared in 1914. The name Zorn's lemma, which has been popularly used, is somewhat misleading; "Zorn's principle" would have been a more suitable name.

In Theorem 2, we actually proved that the axiom of choice implies the Hausdorff maximality principle. In the proof of the next theorem, we shall show the Hausdorff maximality principle implies Zorn's lemma.

Theorem 3. *(Zorn's Lemma).* Let $(A, \leq)$ be a partially ordered set in which every totally ordered subset has an upper bound. Then A has a maximal element.

Proof. By the Hausdorff maximality principle, $(A, \leq)$ has a totally ordered subset B that is maximal with respect to set inclusion $\subseteq$. Let u be an upper bound of B; u exists by hypothesis. We shall prove that the element u is a maximal element of A. If there is an element $x \in A$ such that $x \geq u$, then $B \cup \{x\}$ forms a totally ordered subset of $(A, \leq)$ that contains the maximal totally ordered subset B. Consequently, we must have $B \cup \{x\} = B$, whence $x \leq u$. This proves that u is a maximal element of $(A, \leq)$.

Theorem 3 is a typical existential (as opposed to constructive) proposition; it merely asserts the *existence* of a maximal element in a certain partially ordered set. The proof of Theorem 3 gives no constructive method of finding such a maximal element. A similar comment can be made about Theorem 2 and all the results in the remainder of this chapter.

As an application of Zorn's lemma, we shall establish the following theorem, the proof of which was promised at the end of Section 2, Chapter 6.

Theorem 4. Let A and B be nonempty sets. Then either there is an injection from A to B or there is an injection from B to A.

Proof. We consider the set $\mathscr{X}$ of all pairs (A_α, f_α), where A_α is a subset of A and $f_\alpha : A_\alpha \to B$ is an injection. We define the relation $\leq$ on $\mathscr{X}$ by writing

$$(A_\alpha, f_\alpha) \leq (A_\beta, f_\beta) \qquad \text{if and only if} \qquad A_\alpha \subseteq A_\beta \qquad \text{and} \qquad f_\alpha \subseteq f_\beta$$

This relation is clearly a partial order relation. In order to apply Zorn's lemma, we need to be sure that any totally ordered subset $\mathscr{T} = \{(A_\gamma, f_\gamma) \mid \gamma \in \Gamma\}$ of $\mathscr{X}$ has an upper bound. One natural candidate for an upper bound for $\mathscr{T}$ is $(\bigcup_{\gamma\in\Gamma} A_\gamma, \bigcup_{\gamma\in\Gamma} f_\gamma)$. Denote $A_1 = \bigcup_{\gamma\in\Gamma} A_\gamma$ and $f_1 = \bigcup_{\gamma\in\Gamma} f_\gamma$; then $f_1 : A_1 \to B$ is given by $f_1(x) = f_\gamma(x)$ if $x \in A_\gamma$ and $(A_\gamma, f_\gamma) \in \mathscr{T}$. To prove that $f_1 : A_1 \to B$ is well-defined, suppose that x belongs to another subset A_δ, $\delta \in \Gamma$. Then $(A_\gamma, f_\gamma) \le (A_\delta, f_\delta)$ or $(A_\delta, f_\delta) \le (A_\gamma, f_\gamma)$, and we have $f_\gamma(x) = f_\delta(x)$ in either case. Therefore, $f_1 : A_1 \to B$ is a well-defined function. Next we show that $f_1 : A_1 \to B$ is injective. Suppose that $f_1(x) = f_1(y)$ for some x and y in A_1. Then there exist (A_γ, f_γ) and (A_δ, f_δ) in $\mathscr{T}$ such that $x \in A_\gamma$ and $y \in A_\delta$. As before, either $(A_\gamma, f_\gamma) \le (A_\delta, f_\delta)$ or $(A_\delta, f_\delta) \le (A_\gamma, f_\gamma)$. We may assume that the first alternative is true; then if follows that $f_\delta(x) = f_\delta(y)$, and hence $x = y$, because f_δ is injective. This proves that $f_1 : A_1 \to B$ is injective. Thus, $(A_1, f_1) \ge (A_\gamma, f_\gamma)$ for all $\gamma \in \Gamma$, so that $(\mathscr{X}, \le)$ satisfies the hypothesis of Zorn's lemma. Hence $\mathscr{X}$ has a maximal element, which we denote by $(\tilde{A}, \tilde{f})$. There are two obvious cases:

Case 1. $\tilde{A} = A$.

In this case, $\tilde{f} : A \to B$ is an injection and the theorem holds.

Case 2. $\tilde{A} \ne A$.

Let $x_0 \in A - \tilde{A}$. In this case, we claim that $\tilde{f} : \tilde{A} \to B$ is bijective. For otherwise, there exists an element $y_0 \in B - \tilde{f}(\tilde{A})$. The function $\tilde{\tilde{f}} : \tilde{A} \cup \{x_0\} \to B$, defined by $\tilde{\tilde{f}}(x_0) = y_0$ and $\tilde{\tilde{f}}(x) = \tilde{f}(x)$ for all $x \in \tilde{A}$, is clearly an injection. Thus, $(\tilde{A} \cup \{x_0\}, \tilde{\tilde{f}}) > (\tilde{A}, \tilde{f})$, contradicting the maximality of $(\tilde{A}, \tilde{f})$. This proves that $\tilde{f} : \tilde{A} \to B$ is bijective and consequently, $\tilde{f}^{-1}$ defines an injection of B onto the subset $\tilde{A}$ of A. The proof is now complete.

The following is an immediate consequence of Theorem 4.

Corollary. Let A and B be sets. Then either $\operatorname{card} A \le \operatorname{card} B$ or $\operatorname{card} B \le \operatorname{card} A$.

Thus if m and n are two distinct cardinal numbers, then either $m < n$ or $n < m$.

EXERCISE 7.3

1. Let $(A, \le)$ be a partially ordered set in which every totally ordered subset has a lower bound. Show that A has a minimal element.
2. Prove that Zorn's lemma implies the Hausdorff maximality principle.

3. For students who have studied abstract algebra: prove that every ring with identity has a proper maximal ideal.
4. For students who have studied linear algebra: prove that every vector space has a basis.
5. Prove the corollary to Theorem 4.
6. A partially ordered set $(L, \leq)$ is called a *lattice* if and only if every subset $\{x, y\}$ with two elements has a (unique) least upper bound, denoted by $x \vee y$, and a (unique) greatest lower bound, denoted by $x \wedge y$. Prove that a lattice in which every chain has an upper bound has a unique maximal element.
7. Let $(A, \leq)$ be a partially ordered set in which every totally ordered subset has an upper bound, and let $a \in A$. Then A has a maximal element u such that $u \geq a$.
8. Let B be a set. A set $\mathcal{F}$ of subsets of B is said to have *finite character* provided that $A \in \mathcal{F}$ if and only if every finite subset of A belongs to $\mathcal{F}$. Prove that if $\mathcal{F}$ has finite character, then $(\mathcal{F}, \subseteq)$ has a maximal element.
9. Let A be an arbitrary set with more than one element. Prove that there exists a bijection $f: A \rightarrow A$ such that $f(x) \neq x$, for all $x \in A$.

4. THE WELL-ORDERING PRINCIPLE

We now make another application of Zorn's lemma to prove an astonishing principle in set theory, the Well-Ordering Principle of Ernst Zermelo (1871—1953).

Definition 4. A totally ordered set $(A, \leq)$ is said to be *well ordered* if and only if every nonempty subset B of A contains a (unique) minimal element; that is, if there exists an element $b \in B$ such that $b \leq x$ for every $x \in B$. Such an element b is called the *least element* of B. If $(A, \leq)$ is a well-ordered set then the relation $\leq$ is called a *well-order relation*.

EXAMPLE 5. (a) The set of natural numbers is well ordered under the ordinary "less than or equal" relation. (b) The set of rational numbers under the ordinary "less than or equal" relation is not well ordered.

Let us say that a set A *can be well ordered* if there exists a well-order relation on A.

Theorem 5. (*Well-Ordering Principle*). Every set can be well ordered.

Proof. Let H be an arbitrarly given set which is to be well ordered. Consider the set A^* of all well-ordered sets $(A_0, \leq_0)$, where $A_0 \subseteq H$. We partial-

ly order A^* by writing $(A_0, \leq_0) \leq^* (A_1, \leq_1)$ if and only if
(i) $A_0 \subseteq A_1$
(ii) $x, y \in A_0$ and $x \leq_0 y$ implies $x \leq_1 y$
and
(iii) $x \in A_1 - A_0$ implies $y \leq_1 x$ for all $y \in A_0$.
The reader should verify that this relation $\leq^*$ is really a partial order relation on A^*. In order to apply Zorn's lemma, we show that any totally ordered subset $\mathscr{B}$ of $(A^*, \leq^*)$ has an upper bound. The natural candidate for this upper bound is $(\bigcup_{A \in \mathscr{B}} A, \leq')$,[1] where $x \leq' y$ if and only if both x and y belong to some A_0 such that $(A_0, \leq_0) \in \mathscr{B}$, and $x \leq_0 y$. Obviously, $(\bigcup_{A \in \mathscr{B}} A, \leq')$ is an upper bound for $\mathscr{B}$ if $(\bigcup_{A \in \mathscr{B}} A, \leq')$ belongs to A^*. We shall prove that $(\bigcup_{\in \mathscr{B}} A, \leq')$ is well ordered and hence belongs to A^*. Part of this proof includes the routine verification, which we leave to the reader, that $\leq'$ is a total order relation on $\bigcup_{A \in \mathscr{B}} A$. Let S be a nonempty subset of $\bigcup_{A \in \mathscr{B}} A$. Then there exists $(A_0, \leq_0) \in \mathscr{B}$ such that A_0 intersects S. Since $(A_0, \leq_0)$ is well ordered, $S \cap A_0$ contains a unique least element, say x_0, (of $S \cap A_0$). It follows that for any $y \in S$, there exists $(A_1, \leq_1) \in \mathscr{B}$ such that $(A_0, \leq_0) \leq^* (A_1, \leq_1)$ and $x_0, y \in A_1$; whence, $x_0 \leq_1 y$ and hence $x_0 \leq' y$. Thus, x_0 is the least element for S under $\leq'$, so that $(\bigcup_{A \in \mathscr{B}} A, \leq')$ is well ordered.

By Zorn's lemma, $(A^*, \leq^*)$ has a maximal element $(A_1, \leq_1)$. We claim that $A_1 = H$ and hence $(A, \leq_1)$ is well ordered. For, if $A_1 \neq A$, take any $x_1 \in H - A_1$ and extend $\leq_1$ to $A_1 \cup \{x_1\}$ by defining $x \leq_1 x_1$ for all $x \in A_1$; then $(A_1 \cup \{x_1\}, \leq_1)$ is strictly greater than $(A_1, \leq_1)$ under $\leq^*$, which contradicts the maximality of $(A_1, \leq_1)$. The proof of the well-ordering principle is now complete.

The well-ordering principle is another outstanding example of a nonconstructive proposition. The proof of Theorem 5 gives no indication of how a "well-ordering" of the elements of H is to be accomplished; it merely asserts the existence of a well-order relation. In fact, it is not even known how the set of real numbers can be well ordered.

We have now shown the following chain of implications: axiom of choice $\overset{\text{Th.2}}{\Rightarrow}$ Hausdorff maximality principle $\overset{\text{Th.3}}{\Rightarrow}$ Zorn's lemma $\overset{\text{Th.5}}{\Rightarrow}$ well-ordering principle.

To complete the proof that all these four principles are equivalent, it is sufficient to show that the well-ordering principle implies the axiom of choice.

[1] For simplicity of notation, hereafter the subscript $A \in \mathscr{B}$ will mean $(A, \leq) \in \mathscr{B}$.

Theorem 6. The well-ordering principle implies the axiom of choice.

Proof. Let $\mathscr{I}$ be any nonempty set whose elements are nonempty sets. By the well-ordering principle, there exists a total order relation $\leq$ such that $(\bigcup_{A \in \mathscr{I}} A, \leq)$ is well ordered. Consequently, each set $A \in \mathscr{I}$ contains a least element. Therefore, $f\colon \mathscr{I} \to \bigcup_{A \in \mathscr{I}} A$, defined by $f(A) =$ the least element of A, for all $A \in \mathscr{I}$, is a well-defined choice function. This proves the axiom of choice.

We have thus established the equivalence of the axiom of choice, the Hausdorff maximality principle, Zorn's lemma, and the well-ordering principle. In the remainder of this book we will accept the axiom of choice (and the three other equivalent principles) and use it freely.

EXERCISE 7.4

1. Show that the set of real numbers under the usual "less than or equal" relation of real numbers is not well ordered.
2. Prove directly, without using Theorem 5, that the set of rational numbers can be well ordered.
3. Prove that every subset of a well-ordered set is well ordered under the inherited ordering.
4. Let $(A, \leq)$ be a partially ordered set such that every nonempty subset B contains a lower bound; that is, if there exists $b \in B$ such that $b \leq x$, for all $x \in B$. Prove that the partially ordered set $(A, \leq)$ is totally ordered and hence well ordered.
5. Prove that the relation $\leq^*$ defined in the proof of Theorem 5 is a partial order relation of A^*.
6. Prove that the relation $\leq'$ defined in the proof of Theorem 5 is a total order relation on $\bigcup_{A \in \mathscr{B}} A$.
7. Let $(A, \leq)$ be a totally ordered set. A sequence of elements $a_1, a_2, a_3, \ldots$ in A is said to be *strictly decreasing* if $a_1 > a_2 > a_3 > \cdots$. Prove that a totally ordered set is well ordered if and only if the totally ordered set contains no infinite strictly decreasing sequence.

5. THE PRINCIPLE OF TRANSFINITE INDUCTION

For convenience in expressing the principle of transfinite induction and for the development of ordinal numbers in the next chapter, we now introduce the notion of a segment.

Definition 5. Let $(A, \leq)$ be a totally ordered set. A *segment* of A is a subset S of A such that if $y \in S$, $x \in A$, and $x \leq y$, then $x \in S$. A *proper* segment of A is a segment which is a proper subset of A.

EXAMPLE 6. Let $(A, \leq)$ be a well-ordered set, and let x be an arbitrary element of A. Then the empty set, the set A, and the set $A_x = \{a \in A \mid a < x\}$ are segments of A.

Theorem 7. Let $(A, \leq)$ be a well-ordered set. Then

(a) Any union or intersection of segments of A and all segments of a segment of A are again segments of A.

(b) For each segment S of A, except A itself, there exists an element $x \in A$ such that $S = A_x$, where $A_x = \{a \in A \mid a < x\}$.

Proof. (a) Let $\mathscr{F}$ be a family of segments of A, and let $y \in \bigcup_{S \in \mathscr{F}} S$. If $x \in A$ *and* $x \leq y$, then since y belongs to some segment $S_0 \in \mathscr{F}$ of A, we have $x \in S_0$ and hence $x \in \bigcup_{S \in \mathscr{F}} S$. Therefore, $\bigcup_{S \in \mathscr{F}} S$ is a segment of A.

Similarly, the intersection of the family $\mathscr{F}$ of segments of A is a segment of A (see Problem 3).

Let S be a segment of A and let T be a segment of S. Let $y \in T$, $x \in A$, and $x \leq y$; we are to show that $x \in T$. First, since y belongs to the segment S, we have $x \in S$; then from $y \in T$, $x \in S$, $x \leq y$, and the hypothesis that T is a segment of S, we have $x \in T$. This completes the proof that T is a segment of A.

(b) Let S be a segment of A such that $S \neq A$. Then the nonempty set $A - S$ has a least element, say x. It is an easy exercise for the reader to verify that $S = A_x$.

The following theorem is a prelude to the principle of transfinite induction.

Theorem 8. Let $(A, \leq)$ be a well-ordered set, and let $\mathscr{S}$ be a set of segments of A such that

(1) any union of members of $\mathscr{S}$ belongs to $\mathscr{S}$,

(2) if $A_x \in \mathscr{S}$ then $A_x \cup \{x\} \in \mathscr{S}$.

Then $\mathscr{S}$ contains all segments of A.

Proof. Suppose that there exists an element $x \in A$ such that the segment $A_x \notin \mathscr{S}$. Then, since $(A, \leq)$ is well ordered, the nonempty subset $B = \{x \in A \mid A_x \notin \mathscr{S}\}$ has a least element a. Since $a \in B$, $A_a \notin \mathscr{S}$. If $y \in A$ and $y < a$ then $y \notin B$ and hence $A_y \in \mathscr{S}$. By the hypothesis (1), $\bigcup_{y<a} A_y \in \mathscr{S}$. By

Theorem 7, there exists an element $b \in A$ such that $\bigcup_{y<a} A_y = A_b$. Consequently, $b < a$ (see Problem 9). Now using the hypothesis (2) and Theorem 7, we have

$$A_b \cup \{b\} = A_c \in \mathcal{S} \qquad \text{for some} \qquad c \in A$$

Hence, we have $b < c < a$. Consequently,

$$b \in A_c \subseteq \bigcup_{y<a} A_y = A_b$$

which contradicts the fact that $b \notin A_b$. Therefore, $A_x \in \mathcal{S}$ for all $x \in A$.

It remains to be proved that $A \in \mathcal{S}$. Now by the hypothesis (1) we have $\bigcup_{x \in A} A_x \in \mathcal{S}$. If $\bigcup_{x \in A} A_x = A$, then there is nothing left to be proved. Suppose that $\bigcup_{x \in A} A_x \neq A$; then there exists an element $d \in A$ such that $\bigcup_{x \in A} A_x = A_d$. By the hypothesis (2), $A_d \cup \{d\} \in \mathcal{S}$. It follows that $x \leq d$ for all $x \in A$, and hence $A = A_d \cup \{d\} \in \mathcal{S}$, completing the proof.

Remark. In Theorem 8, the hypothesis (1) implies $\varnothing \in \mathcal{S}$. For, "$\varnothing \in \mathcal{S}$" may be deduced from the hypothesis (1), "any union of members of $\mathcal{S}$ belongs to $\mathcal{S}$," by taking the "empty union" of the members of $\mathcal{S}$ (see Theorem 7(a), Chapter 2). Therefore $\mathcal{S}$ is nonempty.

Theorem 9. *(Principle of Transfinite Induction).* Let $(A, \leq)$ be a well-ordered set. For each $x \in A$, let $p(x)$ be a statement about x. If for each $x \in A$, the hypothesis "$p(y)$ is true for every $y < x$" implies that "$p(x)$ is true," then $p(x)$ is true for every $x \in A$.

Proof. Suppose there is some x such that $p(x)$ is false. Then $B = \{x \in A \mid p(x)$ is false$\}$ is a nonempty subset of A, and hence has a least element x_0. Since $x_0 \in B$, $p(x_0)$ is false. If $y \in A$ and $y < x_0$ then $y \notin B$, and hence $p(y)$ is true. Thus $p(y)$ is true for every $y < x_0$. Thus $p(x_0)$ is true, a contradiction. Therefore, $p(x)$ is true for all x in A.

The next theorem is a typical example of how transfinite induction is used in proving mathematical results. This theorem will be used in the development of the ordinal numbers in Chapter 8. First we need another definition.

Let $(A, \leq)$ and $(B, \leq')$ be well-ordered sets. A function $f: A \to B$ is said to be *increasing* if and only if $a \leq a'$ in $(A, \leq)$ implies $f(a) \leq' f(a')$ in $(B, \leq')$, and to be *strictly increasing* if and only if $a < a'$ in $(A, \leq)$ implies $f(a) <' f(a')$ in $(B, \leq')$.

Theorem 10. Let $(A, \leq)$ and $(B, \leq')$ be well-ordered sets. If $f: A \to B$ is increasing, $f(A)$ is a segment of B, and $g: A \to B$ is strictly increasing, then $f(x) \leq' g(x)$ for all $x \in A$.

Proof. In order to use transfinite induction, for each $x \in A$, let $p(x)$ be the statement "$f(x) \leq' g(x)$." To show that the hypothesis of Theorem 9 is satisfied, we suppose on the contrary that there exists an element $a \in A$ such that $p(x)$ is true for all $x < a$, but $p(a)$ is false. That is $f(x) \leq' g(x)$ for all $x < a$ and $g(a) <' f(a)$. Since g is strictly increasing and f is increasing we have

$$f(x) \leq' g(x) <' g(a) <' f(a) \leq' f(y)$$

for all $x < a$ and for all $y \geq a$. It follows that $g(a) <' f(a)$ and $g(a) \notin f(A)$, contradicting the fact that $f(A)$ is a segment of B. Therefore, for each $x \in A$, if $f(y) \leq' g(y)$ for all $y < x$ then $f(x) \leq' g(x)$. By the principle of transfinite induction, $f(x) \leq' g(x)$ for all $x \in A$.

EXERCISE 7.5

1. Let $\mathbf{N}$ be the set of all natural numbers, and for each $k \in \mathbf{N}$, let $\mathbf{N}_k = \{1, 2, 3, \ldots, k\}$. Find all the segments of $(\mathbf{N}, \leq)$, where $\leq$ denotes the usual "less than or equal" relation for natural numbers.
2. Prove the statement of Example 6.
3. Let $(A, \leq)$ be a well-ordered set. Prove that
   (a) The intersection of an arbitrary family of segments of A is a segment of A.
   (b) For each $x \in A$, $A_x \cup \{x\}$ is a segment of A.
4. Compare the proof of Theorem 7(b).
5. Give a direct proof of Theorem 9. [Hint: use Theorem 8.]
6. Let $(A, \leq)$ and $(B, \leq')$ be well-ordered sets, and let $f: A \to B$ be increasing such that $f(A)$ is a segment of B. Prove that f takes every segment of A to a segment of B.
7. Prove that in a well-ordered set every subset that is bounded above has a (unique) least upper bound
8. Let $(A, \leq)$ be well-ordered. Prove that for each $x \in A$, x is the least upper bound of the segment A_x of A.
9. Prove that $b < a$ in the proof of Theorem 8. [Hint: Consider (i) $b = a$ and (ii) $a < b$.]

6. HISTORICAL REMARKS

It may be worth knowing a brief history of the axiom of choice; in many respects it is similar to that of Euclid's parallel axiom and of the continuum hypothesis (see Section 8, Chapter 6).

As early as the 1880's Georg Cantor had already implicitly used, in the proof of some theorems, reasoning which was essentially equivalent to the axiom of choice; yet he was not aware of using a new powerful axiom. In 1904, Ernst Zermelo (1871-1953) after careful study stated the axiom of choice explicitly and used it to prove the earth-shaking well-ordering theorem (we also called it the well-ordering principle). Because there is no known way of well-ordering even the familiar set of real numbers, despite the assertion of the well-ordering theorem, for a period of at least six years after the appearance of that theorem, many papers appeared that were critical of Zermelo's proof. Most rejected the axiom of choice. Most critics had to admit, however, that if the axiom of choice were accepted they could find no mistake in Zermelo's proof of the well-ordering theorem. Therefore, the rejection of the well-ordering theorem would amount to the rejection of the axiom of choice. There seemed to be only two alternatives:

(a) Accept in principle only constructive but not purely existential results, and consequently, restrict the methods and domains of mathematics to such an extent that, outside of arithmetic, only narrow areas could be investigated.

(b) Accept constructive as well as purely existential results including the axiom of choice, and consequently, solve more problems and expand mathematics.

Before anyone could intelligently decide which alternative to follow, two difficult questions had to be considered:

(1) Is the axiom of choice *independent* of the other axioms, or can it be proved by means of other existing axioms of mathematics?

(2) Is the axiom of choice *consistent* with the classical axioms of mathematics, or can the addition of the axiom of choice to the classical axioms of mathematics result in a contradiction?

As in the case of the continuum hypothesis, many mathematicians spent a great deal of energy trying to answer these two questions. Many years later, in 1938, Kurt Gödel (1906-   ) answered the second question by proving that addition of the axiom of choice to the existing axioms of mathematics would not produce any contradiction.[2] Gödel's contribution

---

[2] K. Gödel, *The Consistency of the Axiom of Choice and of the Generalized Continuum Hypothesis with the Axioms of Set Theorey*, Princetion University Press, Princeton, N.J., 1940, 66 pp. Rev. ed., 1951, 74 pp.

gave the mathematical community and especially the users of the axiom of choice great confidence and comfort. But the search for the answer to the first question went on and on. Finally, in 1963 Paul Cohen solved the question completely. He proved that the axiom of choice was indeed independent of the other existing axioms. In other words, the axiom of choice could not be proved as a theorem using the classical axioms of mathematics.

Today, the axiom of choice has been widely accepted as a new axiom, and it has proved indispensable for modern real analysis, the theory of transfinite cardinal numbers, modern algebra, and for wide areas of topology.

Chapter Eight

Ordinal Numbers and Ordinal Arithmetic

The concept of ordinal numbers is introduced, and the curiosities of ordinal arithmetic are pointed out as the ordinal sum and product are explored. The cardinal numbers are revisited as the initial ordinals, and the Burali-Forti paradox is exhibited.

1. THE CONCEPT OF ORDINAL NUMBERS

Loosely speaking, in finite arithmetic, the cardinal numbers are the "counting" numbers: $1, 2, 3, \ldots$, and the "ordinal numbers" are the "ranking" numbers: first, second, third, and so on. The distinction between the finite cardinal numbers and the finite ordinal numbers is so trivial that the natural numbers may be used to serve in both capacities. But what exactly is an "infinite ordinal number"? Just as a transfinite cardinal number arises from an infinite set, an infinite (= transfinite) ordinal number arises from an infinite well-ordered set. The following definition for well-ordered sets resembles the equipotence relation for general sets.

Definition 1. Two well-ordered sets $(A, \leq)$ and $(B, \leq')$ are said to be *order-isomorphic* if there exists a bijection $f: A \to B$ such that if $a_1, a_2 \in A$ and $a_1 \leq a_2$, then $f(a_1) \leq' f(a_2)$. Such a function $f: A \to B$ is called an *order-isomorphism*.

It follows that if $f: A \to B$ is an order-isomorphism, then so is f^{-1}: $B \to A$, and that if furthermore $g: B \to C$ is an order-isomorphism, then the composition $g \circ f: A \to C$ is an order-isomorphism. If $(A, \leq)$ and $(B, \leq')$

are order-isomorphic, we write $(A, \leq) \approx (B, \leq')$ or simply $A \approx B$. Like the equipotence relation $\sim$, the order-isomorphic "relation" $\approx$ is reflexive, symmetric, and transitive.[1]

In general, we shall consider the *ordinal numbers*, finite or transfinite, as a primitive concept subject to the following rules:

O-1 Each well-ordered set $(A, \leq)$ is assigned an ordinal number, denoted by $\text{ord}(A, \leq)$, and if α is an ordinal number then there is a well-ordered set $(A, \leq)$ such that $\text{ord}(A, \leq) = \alpha$.

O-2 Let $(A, \leq)$ and $(B, \leq')$ be well-ordered sets. Then $\text{ord}(A, \leq) = \text{ord}(B, \leq')$ if and only if $(A, \leq) \approx (B, \leq')$. Since any two finite well-ordered sets having the same number of elements are order-isomorphic (see Problem 1), we shall adopt the following convenient notations:

O-3 $\text{ord}(A, \leq) = 0$ if and only if $A = \varnothing$.

O-4 If $(A, \leq)$ is a well-ordered set such that $A \sim \{1, 2, \ldots, k\}$, for some $k \in \mathbf{N}$, then $\text{ord}(A, \leq) = k$.

The ordinal number of the set $\mathbf{N}$ of natural numbers, with the usual less than of equal relation, is customarily denoted by the *Greek omega*, ω. Thus, $\omega = \text{ord}\{1, 2, 3, \ldots\}$.[2]

A given set has only one cardinal number, but a set may have distinct ordinal numbers under different well-orderings. For example, the set of natural numbers may be well ordered as

$$(\mathbf{N}, \leq) = \{1, 2, 3, \ldots\}$$

and as

$$(\mathbf{N}, \leq') = \{1, 3, 5, \ldots, 2, 4, 6, \ldots\}$$

We leave it to the reader ot verify that these two well-ordered sets are not order-isomorphic. Consequently, $\text{ord}(\mathbf{N}, \leq') \neq \omega$.

---

[1] Here we call $\approx$ a relation in the sense that it is a relation on any given set of well-ordered sets. Cf. Theorem 5 of Chapter 5.

[2] Throughout the remainder of this chapter, the order in which the elements of a set are listed will indicate the ordering of the set. Thus, $\{1, 2, 3, \ldots\}$ denotes the usual well-ordered set of natural numbers, and $\{1, 3, 5, \ldots, 2, 4, 6, \ldots\}$ denotes another ordering of the set $\mathbf{N}$.

EXERCISE 8.1

1. Let A and B *be two equipotent finite sets, and let* $(A, \leq)$ and $(B, \leq')$ be well ordered. Prove that $(A, \leq)$ and $(B, \leq')$ are order-isomorphic.
2. Let $(A, \leq)$ be a totally ordered set which is the union of two subsets B and C such that both B and C are well ordered under the ordering inherited from the ordering $\leq$ of A. Prove that $(A, \leq)$ is well ordered.
3. Let the set $\mathbf{N}$ of natural numbers be well ordered as
$$(\mathbf{N}, \leq) = \{1, 2, 3, \ldots\}$$
and as
$$(\mathbf{N}, \leq') = \{1, 3, 5, \ldots, 2, 4, 6, \ldots\}$$
Prove that $(\mathbf{N}, \leq')$ and $(\mathbf{N}, \leq)$ are not order-isomorphic.
4. Show that $(\mathbf{N}, \leq')$ in Problem 3 above is well ordered.

2. ORDERING OF THE ORDINAL NUMBERS

The rule O-2, given in Section 1, tells us when two ordinal numbers are equal, and when they are not equal. If two ordinal numbers are not equal, we wish to be able to say one is "less" than the other. The following definition is designed just to serve that purpose.

***Definition** 2.* Let α and β be ordinal numbers and let $(A, \leq)$ and $(B, \leq')$ be well-ordered sets such that $\alpha = \operatorname{ord}(A, \leq)$ and $\beta = \operatorname{ord}(B, \leq')$. Then we say that α is *less than or equal* to β, in symbols $\alpha \preccurlyeq \beta$, or $\beta \succcurlyeq \alpha$, if and only if $(A, \leq)$ is order-isomorphic to a segment of $(B, \leq')$. If $\alpha \preccurlyeq \beta$ and $\alpha \neq \beta$, we write $\alpha \prec \beta$ or $\beta \succ \alpha$.

It is clear that Definition 2 is independent of the choice of representing well-ordered sets $(A, \leq)$ and $(B, \leq')$, and that the relation "$\preccurlyeq$" is reflexive and transitive. The antisymmetry of $\preccurlyeq$ is proved in Theorem 2, so that $\preccurlyeq$ is a partial-order "relation". Our eventual goal is to prove that it is a total order "relation."[3]

---

[3] Strictly speaking, "$\preccurlyeq$" is not at relation, because its "domain" is not a set (see Theorem 13). But, we still call it a relation as defined on any given set of ordinal numbers. Cf. Theorem 5 of Chapter 5.

Theorem 1. The only order-isomorphism of a well-ordered set $(A, \leq)$ onto a segment of $(A, \leq)$ is the identity function of A onto A.

Proof. Suppose on the contrary that there exists an order-isomorphism $f: A \to A_a$, for some segment A_a of A. Then $f(a) < a$ and consequently the set $B = \{x \in A \mid f(x) < x\}$ is not empty. Let b denote the least element of B. Then $f(b) < b$, and since an order-isomorphism is strictly increasing, we have $f(f(b)) < f(b)$. This proves that B contains the element $f(b)$, which is less than the least element b of B, a contradiction. The contradiction shows that a well-ordered set cannot be order-isomorphic to any of its proper segments. To complete the proof of the theorem, it remains to be shown that the identity function $I_A : A \to A$ is the only order-isomorphism of A onto A.

Let $g : A \to A$ be an order-isomorphism. Then, since both $g : A \to A$ and the identity function $I_A : A \to A$ are strictly increasing, by using Theorem 10 of Chapter 7 twice, we have $I_A(x) \leq g(x) \leq I_A(x)$ for all $x \in A$. Therefore, $g = I_A$.

Theorem 2. If α and β are ordinal numbers such that $\alpha \preccurlyeq \beta$ and $\beta \preccurlyeq \alpha$, then $\alpha = \beta$.

Proof. Let $(A, \leq)$ and $(B, \leq')$ be well-ordered sets such that $\text{ord}(A, \leq) = \alpha$ and $\text{ord}(B, \leq') = \beta$. By the hypotheses $\alpha \preccurlyeq \beta$ and $\beta \preccurlyeq \alpha$, we have two order-isomorphism

$$f: A \to D \qquad \text{and} \qquad g: B \to C$$

where C and D are segments of A and B, respectively. Consequently, the function $h: A \to C$ given by $h(x) = g(f(x))$, for all $x \in A$, is an order-isomorphism of A to a segment, say E, of C. By Theorem 1, and Theorem 7 of Chapter 7, we must have $E = A$. Consequently, $C = A$ and the order-isomorphism $g: B \to A$ shows that $\alpha = \beta$.

We have now completed the proof that the relation $\preccurlyeq$ for the ordinal numbers is a partial order relation. This property will be strengthened, by the following theorem, to total order.

Theorem 3. For any ordinal numbers α and β, either $\alpha \preccurlyeq \beta$ or $\alpha \succcurlyeq \beta$.

Proof. Let $(A, \leq)$ and $(B, \leq')$ be well-ordered sets such that $\text{ord}(A, \leq) = \alpha$ and $\text{ord}(B, \leq') = \beta$. We shall prove that either A is order-isomorphic to a segment of B, or B is order-isomorphic to a segment to a segment of A. An element a of A will be called *admissible* if the segment A_a of A is order-isomorphic to some segment B_b, $b \in B$, of B. Let M denote the set of all admissible elements of A. It follows from Theorem 1 that for each admissible element a, there exists a *unique* element $b \in B$ such that $A_a \approx B_b$ (see Problem 8). Consequently, there is a well-defined function $f: M \to B$ given by $f(a) = b$ if $A_a \approx B_b$. It is again a consequence of Theorem 1 that the function f is injective. We leave it to the reader to verify that $f: M \to B$ is strictly increasing and that $f(M) = N$ is a segment of B.

Now, suppose that neither A is order-isomorphic to a segment of B nor B is order-isomorphic to a segment of A. Then $A - M$ and $B - N$ are nonempty; let p and q be the least element of $A - M$ and $B - N$, respectively. Then we must have $A_p = M \approx N = B_q$, and hence p is admissible, a contradiction. The proof of Theorem 3 is now complete.

The following theorem is another easy consequence of Theorem 1.

Theorem 4. Let $(A, \leq)$ and $(B, \leq')$ be two well-ordered sets. Then $\text{ord}(A, \leq) \prec \text{ord}(B, \leq')$ if and only if A is order-isomorphic to a proper segment of B.

Proof. Exercise.

We now summarize the results of this section in the following trichotomy theorem.

Theorem 5. Let α and β be any two ordinal numbers. Then exactly one of the following is true:

(a) $\alpha \prec \beta$

(b) $\alpha = \beta$

(c) $\alpha \succ \beta$.

EXERCISE 8.2

1. Show that $\text{ord}\{1, 3, 5, \ldots, 2, 4, 5, \ldots\} \succ \text{ord}\{1, 2, 3, \ldots\}$ (cf. Problem 3,Exercise 8.1).
2. Prove the following assertion that appears in the proof of Theorem 3: For each admissible element a, there exists a *unique* element $b \in B$ such that $A_a \approx B_b$.
3. Show that the set M of admissible elements, in the proof of Theorem 3, is a segment of A.
4. Prove that the function $f: M \to B$, which appeared in the proof of Theorem 3 and is given by $f(a) = b$ if $A_a \approx B_b$, is injective and strictly increasing, and that $f(M)$ is a segment of B.
5. Prove Theorem 4.
6. Let k be any natural number. Prove that

   (a) $\text{ord}\{k, k+1, k+2, \ldots\} = \omega$

(b) ord$\{k, k+1, k+2, \ldots, 0, 1, 2, 3, \ldots, k-1\} \succ \omega$.

7. What is the smallest transfinite ordinal number?
8. From the proof of Theorem 3, show that for each admissible element a, there exists a unique element $b \in B$ such that $A_a \approx B_b$.

3. ADDITION OF ORDINAL NUMBERS

For any two *disjoint* well-ordered sets $(A, \leq)$ and $(B, \leq')$ we shall define the total order relation $\leq^*$ on $A \cup B$ as follows:

(1) If a and b are both in A (or in B), then we write $a \leq^* b$ if and only if $a \leq b$ (or $a \leq' b$).

(2) If $a \in A$ and $b \in B$, we agree that $a \leq^* b$.

It is easy to see that $(A \cup B, \leq^*)$ is a well-ordered set[4] (see Problem 2, Exercise 8.1). Our natural candidate for the "sum," ord$(A, \leq)$ + ord$(B, \leq')$, is ord$(A \cup B, \leq^*)$. In the case where A and B are not disjoint, a little additional work is needed: Form the Cartesian products $A \times \{0\}$, $B \times \{1\}$, and define in $A \times \{0\}$ the well-order relation $\leq_0$ by $(a, 0) \leq_0 (b, 0)$ if and only if $a \leq b$ in $(A, \leq)$. Similarly, define $\leq_1'$ on $B \times \{1\}$ by $(c, 1) \leq_1' (d, 1)$ if and only if $c \leq' d$. It is obvious that $(A \times \{0\}, \leq_0) \approx (A, \leq)$, $(B \times \{1\}, \leq_1') \approx (B, \leq')$, and $A \times \{0\}$ and $B \times \{1\}$ are disjoint. Thus, if A and B are not disjoint, we may use the disjoint sets $(A \times \{0\}, \leq_0)$ and $(B \times \{1\}, \leq_1')$ as reasonable substitutes for $(A, \leq)$ and $(B, \leq')$.

Definition 3. Let α and β be ordinal numbers. The *ordinal sum* of α and β, denoted by $\alpha + \beta$, is the ordinal number ord$(A \cup B, \leq^*)$, where $(A, \leq)$ and $(B, \leq')$ are disjoint well-ordered sets such that ord$(A, \leq) = \alpha$ and ord$(B, \leq') = \beta$.

In Problem 1 the reader is asked to justify that the definition of the ordinal sum $\alpha + \beta$, is independent of the choice of representing well-ordered sets.

If α and β are two finite ordinal numbers, the ordinal sun $\alpha + \beta$ agrees with the usual sum of two nonnegative integers. But, for transfinite ordinal numbers, the properties of the ordinal sum may be very different from the finite case; for example, $\alpha + \beta$ need not be the same as $\beta + \alpha$ (see Example 2, on the next page).

[4] It should be noted that in $(A \cup B, \leq^*)$ any element $a \in A$ *is less than any element* $b \in B$, and that in $(B \cup A, \leq^*)$ any element $b \in B$ is less than any element $a \in A$. Therefore, $(A \cup B, \leq^*)$ and $(B \cup A, \leq^*)$ should be considered distinct.

EXAMPLE 1. Find the ordinal sum $5 + 4$ of the two finite ordinal numbers 5 and 4.

Solution. Since $5 = \text{ord}\{0, 1, 2, 3, 4\}$ and $4 = \text{ord}\{5, 6, 7, 8\}$, we have $5 + 4 = \text{ord}\{0, 1, \ldots, 4, 5, \ldots, 8\} = 9$.

EXAMPLE 2. Let k be any nonzero finite ordinal number. Show that $k + \omega = \omega$ and that $\omega + k \neq \omega$. Thus, the ordinal sum is in general not commutative.

Proof. Since $k = \text{ord}\{0, 1, 2, \ldots, k-1\}$ and $\text{ord}\{k, k+1, \ldots\} = \omega$, we have

$$k + \omega = \text{ord}\{0, 1, 2, \ldots, k-1, k+1, \ldots\} = \omega$$

and

$$\omega + k = \text{ord}\{k, k+1, \ldots, 0, 1, 2, \ldots, k-1\} \succ \omega$$

Thus, $\omega + k \neq \omega$ (see Problem 6(b), Exercise 8.2).

Theorem 6. Let α, β, and γ be any ordinal numbers. Then
(a) $(\alpha + \beta) + \gamma = \alpha + (\beta + \gamma)$ (Associative Law)
(b) $\beta \prec \gamma$ implies $\alpha + \beta \prec \alpha + \gamma$
(c) $\alpha + \beta = \alpha + \gamma$ implies $\beta = \gamma$ (Left Cancellation).

Proof. (a) Exercise.

(b) Let $(A, \leq)$, $(B, \leq')$, and $(C, \leq'')$ be well-ordered sets such that $\text{ord}(A, \leq) = \alpha$, $\text{ord}(B, \leq') = \beta$, $\text{ord}(C, \leq'') = \gamma$, $A \cap B = \varnothing$, and $A \cap C = \varnothing$. Since $\beta \prec \gamma$, by Theorem 4, B is order-isomorphic to some proper segment C_z of C; let $g : B \to C_z$ denote this order-isomorphism. Let the disjoint unions $A \cup B$ and $A \cup C$ be endowed with the well-order relations $\leq^*$ defined in the beginning of this section. Then, under these well-order relations, the function

$$f : A \cup B \to A \cup C_z$$

defined by

$$f(x) = \begin{cases} x & \text{if } x \in A \\ g(x) & \text{if } x \in B \end{cases}$$

is an order-isomorphism of $A \cup B$ onto the proper segment $A \cup C_z$ of $A \cup C$. Therefore, by Theorem 4, $\alpha + \beta \prec \alpha + \gamma$.

(c) Suppose on the contrary that there exist ordinal numbers α, β, and γ such that $\alpha + \beta = \alpha + \gamma$ and $\beta \neq \gamma$. By Theorem 5, we may assume that $\beta \prec \gamma$ (the case $\gamma \prec \beta$ may be handled similarly). Consequently, by part (b) above, we have $\alpha + \beta \prec \alpha + \gamma$, a contradiction.

It is worth mentioning here that although the ordinal sum is left cancellative, it is *not* right cancellative. That is $\beta + \alpha = \gamma + \alpha$ does not necessarily imply that $\beta = \gamma$ (see Problem 4). Simialarly, in contrast to Theorem 6(b), $\beta \prec \gamma$ does not imply $\beta + \alpha \prec \gamma + \alpha$ (see Problems 5 and 6).

EXERCISE 8.3

1. Let $(A_1, \leq_1)$, $(A_2, \leq_2)$, $(B_1, \leq_1')$, and $(B_2, \leq_2')$ be well-ordered sets such that $A_1 \approx A_2$, $B_1 \approx B_2$, $A_1 \cap B_1 = \varnothing$, and $A_2 \cap B_2 = \varnothing$. Prove that $(A_1 \cup B_1, \leq_1^*) \approx (A_2 \cup B_2, \leq_2^*)$.
2. Show that $\alpha + 0 = \alpha = 0 + \alpha$ for any ordinal number α.
3. Prove that the ordinal sum is associative: $(\alpha + \beta) + \gamma = \alpha + (\beta + \alpha)$ for any ordinal numbers α, β, and γ.
4. Show, by giving a counterexample, that the ordinal sum is not right cancellative: there exist ordinal numbers α, β, and γ such that $\beta + \alpha = \gamma + \alpha$, but $\beta \neq \gamma$.
5. Prove the following modification of Theorem 6(b): If α, β, and γ are ordinal numbers such that $\beta \prec \gamma$, then $\beta + \alpha \preccurlyeq \gamma + \alpha$ (see Problem 6, below).
6. Show that, in Problem 5, $\beta \prec \gamma$ does not necessarily imply that $\beta + \alpha \prec \gamma + \alpha$ (cf. Theorem 6(b)).
7. Prove that $\alpha \prec \beta$ if and only if $\alpha + 1 \preccurlyeq \beta$.
8. Prove the following converse of Theorem 6(b): If $\alpha + \beta \prec \alpha + \gamma$ then $\beta \prec \gamma$.
9. Let α and γ be ordinal numbers such that $\alpha \preccurlyeq \gamma$. Prove that there exists a unique β such that $\alpha + \beta = \gamma$. [Such a β may be denoted by $(-\alpha) + \gamma$.]
10. Prove that $\alpha + 1 \succ \alpha$ for any ordinal number.
11. Is there an ordinal number that is greater than all the other ordinal numbers?

4. MULTIPLICATION OF ORDINAL NUMBERS

In cardinal arithmetic, the product of two cardinal numbers $\operatorname{card} A$ and $\operatorname{card} B$ is defined to be $\operatorname{card}(A \times B)$. In any attempt to define the "product" of ordinal numbers $\mathrm{ord}(A, \leq)$ and $\mathrm{ord}(B, \leq')$, we must first decide what well-ordering to impose on the Cartesian product $A \times B$. Our natural preference is the so-called "lexicographic" ordering defined below.

Definition 4. Let $(A, \leq)$ and $(B, \leq')$ be well-ordered sets. Then the *lexicographic* ordering $\leq^*$ of $A \times B$ is defined by: (a) If $a < x$ then $(a, b) \leq^* (x, y)$ for any b and y in B. (b) If $a = x$ and $b \leq' y$ then $(a, b) \leq^* (x, y)$.

Theorem 7. Let $(A, \leq)$ and $(B, \leq')$ be well-ordered sets. Then the lexicographic ordering $\leq^*$ of $A \times B$ is a well-order relation on $A \times B$.

Proof. It is clear that $\leq^*$ is a total order relation on $A \times B$. To show that the total order relation is a well-ordering, let S be an arbitrary nonempty subset of $A \times B$. We shall prove that S contains a least element. First, we observe that the set

$$p_A(S) = \{x \in A \mid (x, y) \in S \text{ for some } y \in B\}$$

is a nonempty subset of the well-ordered set A, and hence contains a least element, say a. Then consider the set

$$\{y \in B \mid (a, y) \in S\}$$

which is a nonempty subset of the well-ordered set B and hence contains a least element, say b. It is now quite obvious that the element (a, b) in S is the least element of S. Therefore, $(A \times B, \leq^*)$ is a well-ordered set.

Theorem 8. Let $(A_1, \leq_1)$, $(A_2, \leq_2)$, $(B_1, \leq_1')$, and $(B_2, \leq_2')$ be well-ordered sets such that $(A_1, \leq_1) \approx (A_2, \leq_2)$ and $(B_1, \leq_1') \approx (B_2, \leq_2')$. Then

$$(A_1 \times B_1, \leq_1^*) \approx (A_2 \times B_2, \leq_2^*)$$

Proof. Let $f: A_1 \to A_2$ and $g: B_1 \to B_2$ be order-isomorphisms. We leave it as an exercise for the reader to show that the function

$$f \times g: A_1 \times B_1 \to A_2 \times B_2$$

defined by $(f \times g)(x, y) = (f(x), g(y))$ for all $(x, y) \in A_1 \times B_1$ is an order-isomorphism.

Theorems 7 and 8 permit us to define the product of two ordinal numbers.

Definition 5. For any ordinal numbers α and β, the *ordinal product* $\beta\alpha$ is defined by $\beta\alpha = \text{ord}(A \times B, \leq^*)$, where $(A, \leq)$ and $(B, \leq')$ are well-ordered sets such that $\text{ord}(A, \leq) = \alpha$, $\text{ord}(B, \leq') = \beta$, and $\leq^*$ is the lexicographic ordering of $A \times B$.

Note that, according to Definition 5, $\text{ord}(A \times B, \leq^*)$ is $\beta\alpha$, *not* $\alpha\beta$. The ordinal product is not commutative.

EXAMPLE 3. Compare the ordinal products 2ω and $\omega 2$.

Solution. Let $(A, \leq) = \{1, 2, 3, \ldots\}$ and $(B, \leq') = \{0, 1\}$ so that $\text{ord}(A, \leq) = \omega$ and $\text{ord}(B, \leq') = 2$. Let $A \times B$ be endowed with the lexicographic ordering so that $2\omega = \text{ord}(A \times B, \leq^*)$. Then the function $f: A \times B \to A$ given by

$$f(j, k) = \begin{cases} 2j - 1 & \text{if} \quad k = 0 \\ 2j & \text{if} \quad k = 1 \end{cases}$$

is an order-isomorphism. Therefore, $2\omega = \omega$.

Next, impose the lexicographic ordering $\leq'^*$ on $B \times A$ so that $\omega 2 = \text{ord}(B \times A, \leq'^*)$. We leave it to the reader to verify that $\text{ord}(B \times A, \leq'^*) = \omega + \omega$.

Therefore, $2\omega \neq \omega 2$.

Theorem 9. Let α, β, and γ be any ordinal numbers. Then

(a) $(\gamma\beta)\alpha = \gamma(\beta\alpha)$ (Associative Law)

(b) $\gamma(\alpha + \beta) = \gamma\alpha + \gamma\beta$ (Left Distributive Law).

Proof. Exercise.

Theorem 10. Let α, β, and γ be any ordinal numbers such that $\gamma \succ 0$. Then

(a) $\alpha \prec \beta$ implies $\gamma\alpha \prec \gamma\beta$

(b) $\gamma\alpha = \gamma\beta$ implies $\alpha = \beta$ (Left Cancellation).

Proof. (a) Let $(A, \leq)$, $(B, \leq')$, and $(C, \leq'')$ be well-ordered sets such that $\operatorname{ord}(A, \leq) = \alpha$, $\operatorname{ord}(B, \leq') = \beta$, and $\operatorname{ord}(C, \leq'') = \gamma$. If $\alpha < \beta$, then there exists an element $p \in B$ such that $A \approx B_p$. Let q denote the least element of C, and let $A \times C$ and $B \times C$ be endowed with the lexicographic orderings. It follows that $A \times C$ is isomorphic to $B_p \times C$ under the inherited ordering from $B \times C$, and that $B_p \times C$ is the segment

$$(B \times C)_{(p,q)} = \{(x, y) \in B \times C \mid (x, y) <'^{*} (p, q)\}$$

of $B \times C$. This proves that $\gamma\alpha \prec \gamma\beta$.

(b) Suppose on the contrary that there exist ordinal numbers α, β, and $\gamma \succ 0$ such that $\gamma\alpha = \gamma\beta$ and $\alpha \neq \beta$. We assume that $\alpha \prec \beta$ (the case where $\beta \prec \alpha$ may be treated similarly). Then, by part (a) of this theorem we have $\gamma\alpha \succ \gamma\beta$, a contradiction.

It should be noted that although ordinal multiplication is left distributive and left cancellative, it is neither right distributive nor right cancellative.(See Problem 4 and 9.)

EXERCISE 8.4

1. Complete the proof of Theorem 8.
2. Show that
   (a) $0\alpha = 0 = \alpha 0$ and $1\alpha = \alpha = \alpha 1$ for any ordinal number α
   (b) $\alpha\beta = 0$ if and only if either $\alpha = 0$ or $\beta = 0$.
3. Verify that
   (a) $\omega 2 = \omega + \omega$
   (b) $k\omega = \omega$ for any finite ordinal $k \neq 0$.
4. Show by giving a counterexample, that ordinal multiplication is not *right distributive*, that is, $(\beta + \gamma)\alpha = \beta\alpha + \gamma\alpha$ need not be true.
5. Prove that ordinal multiplication is left distributive, that is, $\gamma(\alpha + \beta) = \gamma\alpha + \gamma\beta$ for all ordinal numbers α, β, and γ.
6. Prove that ordinal multiplication is associative: $(\gamma\beta)\alpha = \gamma(\beta\alpha)$ for all ordinal numbers α, β, and γ.
7. Prove the following modification of Theorem 10(a): if α, β, and γ are such that $\alpha \prec \beta$ and $\gamma \succ 0$, then $\alpha\gamma \preccurlyeq \beta\gamma$. (See Problem 8, below.)
8. Show that, in Problem 7 above, $\alpha \prec \beta$ and $\gamma \succ 0$ do not necessarily imply that $\alpha\gamma \prec \beta\gamma$ (cf. Theorem 10(a)).
9. Show, by giving a counterexample, that ordinal multiplication is not right cancellative: there exist ordinal numbers α, β, and $\gamma \succ 0$ such that $\alpha\gamma = \beta\gamma$, but $\alpha \neq \beta$.
10. Prove the following converse of Theorem 10(a): If α, β, and γ are ordinal numbers such that $\gamma\alpha \prec \gamma\beta$ and $\gamma \succ 0$, then $\alpha \prec \beta$.

5. CONCLUSION

It may seem natural now to explore the exponentiation of the ordinal numbers. We choose, however, to leave this to more advanced books. Instead, we shall first re-examine the ordering of the ordinal numbers, and then revisit the cardinal numbers from the ordinal-number-theoretical point of view.

Theorem 11. Let α be an arbitrary ordinal number. Then the set of all ordinal numbers β such that $\beta \prec \alpha$ is a well-ordered set whose ordinal number is α.

Proof. Let $(A, \leq)$ be a well-ordered set whose ordinal number is α. For any ordinal number β with $\beta \prec \alpha$ and any well-ordered set $(B, \leq')$ with $\operatorname{ord}(B, \leq') = \beta$, β is order-isomorphic to a proper segment A_b, $b \in A$, of A. It follows from Theorem 1 that the element $b \in A$ is uniquely determined by the ordinal number β. Consequently, there is a well-defined function

$$f: \{\beta \mid \beta \text{ is an ordinal number with } \beta \prec \alpha\} \to A$$

given by $f(\beta) = b$ if $\beta = \operatorname{ord}(B, \leq')$ and $B \approx A_b$. It is quite routine, and therefore left to the reader, to show that the function f is an order-isomorphism. Therefore, the set $\{\beta \mid \beta \prec \alpha\}$ is well ordered and

$$\operatorname{ord}\{\beta \mid \beta \prec \alpha\} = \alpha$$

In view of Theorem 11, it may be convenient to identify the ordinal number α with the set $\{\beta \mid \beta \prec \alpha\}$ and thus to regard each ordinal number as a well-ordered set (of ordinal numbers). For example:

$$0 \equiv \varnothing$$
$$1 \equiv \{0\}$$
$$2 \equiv \{0, 1\}$$
$$3 \equiv \{0, 1, 2\}$$
$$\cdot$$
$$\cdot$$
$$\cdot$$
$$\omega \equiv \{0, 1, 2, \ldots\}$$
$$\omega + 1 \equiv \{0, 1, 2, \ldots, \omega\}$$
$$\omega + 2 \equiv \{0, 1, 2, \ldots, \omega, \omega + 1\}$$
$$\cdot$$
$$\cdot$$
$$\cdot$$
$$\omega 2 \equiv \{0, 1, \ldots, \omega, \omega + 1, \ldots\}$$
$$\omega 2 + 1 \equiv \{0, 1, \ldots, \omega, \omega + 1, \ldots, \omega 2\}$$
$$\cdot$$
$$\cdot$$
$$\cdot$$

Theorem 12. Any set of ordinal numbers is well ordered.

Proof. Suppose on the contrary that there exists a set A of ordinal numbers that is not well ordered. Then there is a subset B of A that does not have a least element. Consequently, the set B contains a strictly decreasing infinite sequence of ordinal numbers $\alpha_1 \succ \alpha_2 \succ \alpha_3 \succ \cdots$. This sequence is contained in $\{\beta \mid \beta \prec \alpha_1\}$, and hence, contradicting Theorem 11, the set $\{\beta \mid \beta \prec \alpha_1\}$ is not well ordered. This completes the proof of the theorem.

Viewing ordinal numbers as sets, let us consider, for example, the set $\mathcal{N}$ of all ordinal numbers α (α's are sets!) with α equipontent to the set $\mathbf{N}$ of natural numbers. This set includes:

$$\omega, \omega + 1, \ldots, \omega 2, \omega 2 + 1, \ldots, \omega 3, \omega 4, \ldots, \omega^2, \omega^3, \ldots, \omega^\omega, \ldots, \omega^{(\omega^\omega)}, \ldots, \omega^{\omega^{\cdot^{\cdot^{\cdot}}}}, \ldots.$$

According to Theorem 12, the set $\mathcal{N}$ is well ordered, and hence there is a unique least ordinal number that is equipotent to $\mathbf{N}$. It is called the *initial ordinal* for the set $\mathbf{N}$. (It is not difficult to see that the initial ordinal for $\mathbf{N}$ is ω.) In general, by the well-ordering principle and by Theorem 12, every set X has a (unique) initial ordinal. The initial ordinals satisfy the guiding rules C-1 to C-4 for the cardinal numbers (see Section 1, Chapter 6). This suggests an alternate approach to the cardinal numbers. As a matter of fact, some authors introduce the ordinal numbers first and then define the cardinal number of a set as the initial ordinal of that set. There is a logical advantage to treating the cardinal numbers as the initial ordinals, but from a pedagogical and practical point of view we chose to explore the cardinal numbers before the ordinal numbers.

In Theorem 12, we have carefully avoided the phrase "the set of all ordinal numbers." As in the case of Russell's paradox, the assumption of the existence of this set leads to a contradiction, known as the Burali-Forti paradox

Theorem 13. There does not exist a set of all ordinal numbers.

Proof. Suppose on the contrary that there is a set S of all ordinal numbers. By Theorem 12, S is well ordered. The ordinal number of S, denoted by σ, must be a member of S. It follows from Theorems 11 and 4 that

$$\sigma = \operatorname{ord}\{\beta \in S \mid \beta \prec s\} = \operatorname{ord} S_\sigma \prec \operatorname{ord} S = \sigma$$

which is a contradiction.

Alternative Proof. Let S and σ be as defined above. Then σ is the largest ordinal number, which contradicts the fact that $\sigma \prec \sigma + 1$ (see Problem 10, Exercise 8.3).

EXERCISE 8.5

1. Prove that the function $f: \{\beta \mid \beta \prec \alpha\} \to A$, given in the proof of Theorem 11, such that $f(\beta) = b$ if $\beta = \text{ord}(B, \leq')$ for some well-ordered set $(B, \leq')$ and $B \approx A_b$ is an order-isomorphism.
2. Prove that the initial ordinal for the set $\mathbf{N}$ is ω.
3. Show that every set has a unique initial ordinal.
4. Let X and Y be sets. Prove that $X \sim Y$ if and only if X and Y have the same initial ordinal.
5. Let X and Y be sets. Prove that $\text{card}\, X < \text{card}\, Y$ if and only if the initial ordinal of X is less than the initial ordinal of Y.

Appendix

The Peano Axioms for Natural Numbers

The system of natural numbers 1, 2, 3, . . . serves as the starting point for constructing the system of integers, the system of rational numbers, the system of residue classes modulo an integer, etc. What really are natural numbers? Guiseppi Peano (1858-1932) answered this question with five axioms, the Peano axiom for natural numbers.[1]

The Peano Axioms for Natural Numbers. There exists a set **N** of elements, called the natural numbers, satisfying the following axioms:

1. there exists a special element in **N** which we denote by 1;
2. for each element $n \in \mathbf{N}$, there exists a unique element n^+, called the *successor* of n, in **N**;
3. for all $n \in \mathbf{N}$, $n^+ \neq 1$;
4. if $n, m \in \mathbf{N}$ and $n^+ = m^+$, then $n = m$;
5. if **P** is a subset of **N** such that $1 \in \mathbf{P}$, and $n \in \mathbf{P}$ implies $n^+ \in \mathbf{P}$, then $\mathbf{P} = \mathbf{N}$.

It may be more convenient to write $n + 1$ for the successor n^+ of n, and we write $2 = 1 + 1$, $3 = 2 + 1$, $4 = 3 + 1$, etc.

Axiom 5 is the basis of mathematical induction. Addition of natural numbers is defined inductively by:

---

[1] See G. Peano, *Arithmetices Principia*, Bocca, Turin, 1889.

(1) $$1 + y = y^+$$

(2) $$x^+ + y = (x + y)^+$$

Using (1), (2), and mathematical induction, the following basic properties of addition of natural numbers can be proved.

(3) $x + (y + z) = (x + y) + z$ (associative)

(4) $x + y = y + x$ (commutative law)

(5) $x + z = y + z$ implies $x = y$ (cancellation law)

We shall prove only (3) here and leave (4) and (5) to the reader as exercises.

Proof (3). Let y and z be any two arbitrary members of **N** that are fixed throughout this proof. In order to prove (3) by mathematical induction, we first show that

$$1 + (y + z) = (1 + y) + z$$

This is accomplished by the following step-by-step derivation:

$$\begin{aligned} 1 + (y + z) &= (y + z)^+ && \text{by (1)} \\ &= y^+ + z && \text{by (2)} \\ &= (1 + y) + z && \text{by (1)} \end{aligned}$$

Thus, the associative law (3) is valid for $x = 1$ and the first requirement for mathematical induction is satisfied. Our next task is to show that the validity of

$$k + (y + z) = (k + y) + z \qquad \text{(induction hypothesis)}$$

implies the validity of

$$k^+ + (y + z) = (k^+ + y) + z$$

We have

$$\begin{aligned} k^+ + (y + z) &= [k + (y + z)]^+ && \text{by (2)} \\ &= [(k + y) + z]^+ && \text{by induction hypothesis} \\ &= (k + y)^+ + z && \text{by (2)} \\ &= (k^+ + y) + z && \text{by (2)} \end{aligned}$$

which was to be proved. The proof of (3) is now complete, by mathematical induction.

Multiplication of natural numbers is defined inductively by:

(6) $$1y = y$$

(7) $$x^+y = xy + y$$

The following basic properties of multiplication can be derived from (6) and (7) by mathematical induction:

(8) $x(yz) = (xy)z$ (associative law)

(9) $xy = yz$ (commutative law)

(10) $xz = yz$ implies $x = y$ (cancellation law)

(11) $x(y+z) = xy + xz$ (distributive law)

Properties (8), (9), and (10) for multiplication resemble properties (3), (4), and (5), respectively, for addition. The proofs for (8), (9), and (10) are similar to the proofs for (3), (4), and (5), and hence the proofs of these are again left to the reader. We prove the distributive law here.

Proof (11). Let y and z be any two arbitrary members of **N** that are fixed throughout the proof. We shall prove (11) by mathematical induction on x. For $x = 1$, by (6), it is clearly true that

$$1(y+z) = 1y + 1z$$

Thus, the first part of mathematical induction is satisfied. To complete the proof, we assume that

$$k(y+z) = ky + kz$$

is true for some $k \in \mathbf{N}$. Then

$$\begin{aligned}
k^+(y+z) &= k(y+z) + (y+z) && \text{by (7)}\\
&= (ky + kz) + (y+z) && \text{by induction hypothesis}\\
&= ky + [kz + (y+z)] && \text{by (3)}\\
&= ky + [(kz + y) + z] && \text{by (3)}\\
&= ky + [(y + kz) + z] && \text{by (4)}\\
&= ky + [y + (kz + z)] && \text{by (3)}\\
&= (ky + y) + (kz + z) && \text{by (3)}\\
&= k^+y + k^+z && \text{by (7)}
\end{aligned}$$

Therefore, by mathematical induction, (11) is valid.

The concept of order in **N** can be introduced as follows:

(12) $x > y$ if and only if there exists $z \in \mathbf{N}$ such that $x = y + z$.

From this definition of order, it follows that

(13) $x > y$ and $y > z$ imply $x > z$ (transitive law)

and

(14) for any x and y in **N**, one and only one of the following holds:

$$x > y, \ x = y, \ y > x \text{ (trichotomy law)}$$

In fact, all known properties of natural numbers may be derived from the Peano axioms. In concluding this appendix, we state the following theorem, which asserts that the natural number system is unique.

Theorem. Let N and N' be two sets satisfying the Peano axioms 1 through 5. Then there exists a one-to-one correspondence (called an isomorphism) $f: N \to N'$ such that $f(1) = 1'$ and $f(n+1) = f(n) + 1'$, where $1'$ denotes the special element of N' satisfying 1—5.

Answers

Answers to Selected Problems

EXERCISE 1.1

1. (S) 2. (S) 3. (N) 4. (S) 5. (N) 6. (N)
7. (N) 8. (N) 9. (S) 10. (S) 20. 8, 16, 2^n

21.

p	q	r	
T	T	T	
T	T	F	
T	F	T	
T	F	F	
F	T	T	
F	T	F	
F	F	T	
F	F	F	

EXERCISE 1.2

13. yes 14. yes 15. (4) and (5), (7) and (8), (9) and (10), (11) and (12)
19. yes 20. yes 21. $\sim p \wedge \sim q$

EXERCISE 1.3

8. Neither this function has a derivative nor am I stupid.

10. (a) $\sim(p_1 \wedge p_2 \wedge p_3 \wedge \cdots \wedge p_n) \equiv \sim p_1 \vee \sim p_2 \vee \sim p_3 \vee \cdots \vee \sim p_n$
(b) $\sim(p_1 \vee p_2 \vee p_3 \vee \cdots \vee p_n) \equiv \sim p_1 \wedge \sim p_2 \wedge \sim p_3 \wedge \cdots \wedge \sim p_n$

14. yes 18. $p \wedge q \wedge r$ 19. $\sim p \wedge \sim q \wedge \sim r$

EXERCISE 1.4

3. From the truth table:

	p	$\wedge$	$\sim q$	$\rightarrow$	c	$\leftrightarrow$	p	$\rightarrow$	q
	T	F	F	T	F	T	T	T	T
	T	T	T	F	F	T	T	F	F
	F	F	F	T	F	T	F	T	T
	F	F	T	T	F	T	F	T	F
Step	1	3	2	4	1	5	1	2	1

we conclude that $p \wedge \sim q \rightarrow c$ and $p \rightarrow q$ are equivalent.

EXERCISE 1.5

1. $p \wedge (p \rightarrow q) \equiv p \wedge \sim(p \wedge \sim q)$ Def. of $\rightarrow$
$\equiv p \wedge (\sim p \vee q)$ De M., D.N.
$\equiv (p \wedge \sim p) \vee (p \wedge q)$ Dist.
$\equiv c \vee (p \wedge q)$ $p \wedge \sim p \equiv c$
$\equiv p \wedge q$ Com., Th. 7(b)
$\Rightarrow q$ Simp.

3. $(p \wedge \sim q \rightarrow c) \Leftrightarrow \sim[(p \wedge \sim q) \wedge \sim c]$ Def. of $\rightarrow$
$\Leftrightarrow \sim(p \wedge \sim q) \vee c$ De M., D.N.
$\Leftrightarrow \sim(p \wedge \sim q)$ Th. 7(b)
$\Leftrightarrow (p \rightarrow q)$ Def. of $\rightarrow$

4. $(p \vee q \vee r) \wedge \sim r \wedge \sim q \Leftrightarrow [(p \vee q \vee r) \wedge \sim r] \wedge \sim q$ Assoc.,
$\Rightarrow (p \vee q) \wedge \sim q$ Example 6
$\Rightarrow p$ Example 6

5. $c \Leftrightarrow p \wedge \sim p$ Def.
$\Rightarrow p$ Simp.

7. $(p \vee q \rightarrow q) \Leftrightarrow \sim[(p \vee q) \wedge \sim q]$ Def. of $\rightarrow$
$\Leftrightarrow \sim(p \vee q) \vee q$ De M., D.N.
$\Leftrightarrow q \vee (\sim p \wedge \sim q)$ De M., Com.

$\Leftrightarrow (q \vee \sim p) \wedge (q \vee \sim q)$	Dist.
$\Leftrightarrow (\sim p \vee q) \wedge t$	Com., $q \vee \sim q \equiv t$
$\Leftrightarrow \sim p \vee q$	Th. 7(a)
$\Leftrightarrow \sim (p \wedge \sim q)$	De M., D.N.
$\Leftrightarrow (p \rightarrow q)$	Def. of $\rightarrow$

9.

$(p \rightarrow r) \wedge (q \rightarrow r) \Leftrightarrow (\sim p \vee r) \wedge (\sim q \vee r)$	Prob. 8
$\Leftrightarrow (r \vee \sim p) \wedge (r \vee \sim q)$	Com.
$\Leftrightarrow r \vee (\sim p \wedge \sim q)$	Dist.
$\Leftrightarrow \sim (p \vee q) \vee r$	De M., Com.
$\Leftrightarrow (p \vee q \rightarrow r)$	Prob. 8

11.

$(p \rightarrow q) \wedge (p \rightarrow \sim q) \Leftrightarrow (\sim p \vee q) \wedge (\sim p \vee \sim q)$	Prob. 8
$\Leftrightarrow \sim p \vee (q \wedge \sim q)$	Dist.
$\Leftrightarrow \sim p \vee c$	$q \wedge \sim q \equiv c$
$\Leftrightarrow \sim p$	Th. 7(b)

12.

$(p \rightarrow q) \vee (p \rightarrow r) \Leftrightarrow (\sim p \vee q) \vee (\sim p \vee r)$	Prob. 8
$\Leftrightarrow (\sim p \vee \sim p) \vee (q \vee r)$	Assoc., Com.
$\Leftrightarrow \sim p \vee (q \vee r)$	Idemp.
$\Leftrightarrow (p \rightarrow q \vee r)$	Prob. 8

16.

$[p \rightarrow (q \rightarrow r)] \wedge (p \rightarrow q)$	
$\Leftrightarrow [\sim p \vee (q \rightarrow r)] \wedge (\sim p \vee q)$	Def. 4, De M.
$\Leftrightarrow \sim p \vee [(q \rightarrow r) \wedge q]$	Dist.
$\Rightarrow \sim p \vee r$	M.P., Prob. 5, Ex. 1.4

20.

$(p \rightarrow q) \wedge (r \rightarrow s) \wedge (\sim q \vee \sim s)$	
$\Leftrightarrow (p \rightarrow q) \wedge (r \rightarrow s) \wedge (q \rightarrow \sim s)$	$\sim q \vee \sim s \equiv q \rightarrow \sim s$
$\Leftrightarrow [(p \rightarrow q) \wedge (q \rightarrow \sim s)] \wedge (r \rightarrow s)$	Com., Assoc.
$\Rightarrow (p \rightarrow \sim s) \wedge (\sim s \rightarrow \sim r)$	Trans., Contrap.
$\Rightarrow p \rightarrow \sim r$	Trans.
$\Leftrightarrow \sim p \vee \sim r$	Def. 4, De M.

EXERCISE 1.6

4. Substituting $\sim p(x)$ for $q(x)$ in $\sim [(\forall x)(q(x))] \equiv (\exists x)(\sim q(x))$, we have

$$\sim [(\forall x)(\sim p(x))] \equiv (\exists x)(p(x))$$

Negating both sides of the above equivalence and interchanging the left-and the right-hand sides, we conclude

$$\sim [(\exists x)(p(x))] \equiv (\forall x)(\sim p(x))$$

EXERCISE 1.7

5. (a) Direct Proof:

1.	$A \vee (B \wedge C)$	
2.	$B \rightarrow D$	
3.	$C \rightarrow E$	
4.	$D \wedge E \rightarrow A \vee F$	
5.	$\sim A \;/\; \therefore\; F$	
6.	$B \wedge C$	1, 5, D.S.
7.	$B \wedge C \rightarrow D \wedge E$	2, 3, C.D.
8.	$D \wedge E$	7, 6, M.P.
9.	F	4, 8, M.P.

(b) Indirect Proof:

6.	$\sim F$	I.P.
7.	$\sim (D \wedge E)$	4, 6, M.T.
8.	$B \wedge C$	1, 5, D.S.
9.	$B \wedge C \rightarrow D \wedge E$	2, 3, C.D.
10.	$D \wedge E$	9, 8, M.P.
11.	$(D \wedge E) \wedge \sim (D \wedge E)$	10, 7, Conj.

6. (a) Direct Proof:

1.	$B \vee (C \rightarrow E)$	
2.	$B \rightarrow D$	
3.	$\sim D \rightarrow (E \rightarrow A)$	
4.	$\sim D \;/\; \therefore\; C \rightarrow A$	
5.	$\sim B$	2, 4, M.T.
6.	$C \rightarrow E$	1, 5, D.S.
7.	$E \rightarrow A$	3, 4, M.P.
8.	$C \rightarrow A$	6, 7, Trans.

(b) Indirect Proof:

5.	$\sim (C \rightarrow A)$	I.P.
6.	$C \wedge \sim A$	5, Def. 4, D.N.
7.	$\sim A$	6, Simp.
8.	$E \rightarrow A$	3, 4, M.P.
9.	$\sim E$	7, 8, M.T.
10.	$\sim B$	2, 4, M.T.
11.	$C \rightarrow E$	1, 10, D.S.
12.	C	6, Simp.
13.	E	11, 12, M.P.
14.	$E \wedge \sim E$	13, 9, Conj.

10. (a) Direct Proof:

1. $A \wedge B \rightarrow C$
2. $(A \rightarrow C) \rightarrow D$

3.	$\sim B \vee E \,/\, \therefore\ B \to D \wedge E$	
4.	$B \to (A \to C)$	1, Com., Expotortation
5.	$B \to D$	4, 2, Trans.
6.	$B \to E$	3, Def. 4, De M.
7.	$B \wedge B \to D \wedge E$	5, 6, C.D.
8.	$B \to D \wedge E$	7, Idemp.

11.

1.	$P \wedge C \to R$	
2.	$R \to G$	
3.	$H \to \sim I$	
4.	H	
5.	$\sim G \vee I \,/\, \therefore\ \sim (P \wedge C)$	
6.	$\sim I$	3, 4, M.P.
7.	$\sim G$	5, 6, D.S.
8.	$\sim R$	2, 7, M.T.
9.	$\sim (P \wedge C)$	1, 8, M.T.

13.

1.	$W \vee H \to L \wedge S$	
2.	$\sim S \,/\, \therefore\ \sim H$	
3.	$\sim S \vee \sim L$	2, Add.
4.	$\sim (L \wedge S)$	3, Com., De M.
5.	$\sim (W \vee H)$	1, 4, M.T.
6.	$\sim W \wedge \sim H$	5, De M.
7.	$\sim H$	6, Simp.

14.

1.	$E \wedge S \to G$	
2.	$G \to H$	
3.	$\sim H \,/\, \therefore\ \sim E \vee \sim S$	
4.	$\sim G$	2, 3, M.T.
5.	$\sim (E \wedge S)$	1, 4, M.T.
6.	$\sim E \vee \sim S$	5, De M.

15.

1.	$P \to \sim B$	
2.	$F \to T$	
3.	$O \to \sim S$	
4.	$\sim B \to F$	
5.	$G \to P$	
6.	$\sim S \to \sim T \,/\, \therefore\ O \to \sim G$	
7.	$O \to \sim T$	3, 6, Trans.
8.	$G \to \sim B$	5, 1, Trans.

9.	$G \rightarrow F$	8, 4, Trans.
10.	$G \rightarrow T$	9, 2, Trans.
11.	$\sim T \rightarrow \sim G$	10, Contrap.
12.	$O \rightarrow \sim G$	7, 11, Trans.

EXERCISE 1.8

1. We prove this theorem by mathematical induction on n. The theorem is easily verified for $n = 1$. The induction hypothesis is that k is an integer such that for all r with $0 \leq r \leq k$,

$$C(k, r) = \frac{k!}{r!(k-r)!}$$

Now consider $C(k+1, r)$. By Definition 7 and the induction hypothesis we have

$$C(k+1, r) = C(k, r) + C(k, r-1)$$

$$= \frac{k!}{r!(k-r)!} + \frac{k!}{(r-1)!(k-r+1)!}$$

$$= \frac{k!(k-r+1)}{r!(k-r)!(k-r+1)} + \frac{k!r}{r(r-1)!(k-r+1)!}$$

$$= \frac{k!(k+1)}{r!(k-r+1)!}$$

$$= \frac{(k+1)!}{r!(k+1-r)!}$$

which shows that the theorem is true for $k+1$ if it is true for k. The proof is now complete by the principle of mathematical induction.

2. Theorem 8 may be used for this problem. If mathematical induction is to be used, prove first the following lemma by mathematical induction.
Lemma. If n is a nonnegative integer and r is either less than zero or greater than n, then $C(n, r) = 0$.

EXERCISE 2.1

2. $D \subseteq A \subseteq B \subseteq C$

5. Suppose that $A \subseteq \varnothing$; then for any element x, $(x \in A) \to (x \in \varnothing)$ is a true statement. Since $x \in \varnothing$ is false, in order that the conditional statement $(x \in A) \to (x \in \varnothing)$ be true, $x \in A$ must be false for all elements x. Thus, we must have $A = \varnothing$.

6. (a) We verify the validity of the following argument:

1.	$(x \in A) \to (x \in B)$,	for all elements x	(Hyp. $A \subset B$)
2.	$(y \in B) \wedge (y \notin A)$,	for some element y	
3.	$(x \in B) \to (x \in C)$,	for all elements x	(Hyp. $B \subseteq C$)
		$/ \therefore A \subset C$	(Concl.)
4.	$(x \in A) \to (x \in C)$,	for all elements x	1, 3, Trans.
5.	$y \in B$		2, Simp.
6.	$y \notin A$		2, Simp.
7.	$y \in C$		3, 5, M.P.
8.	$(y \in C) \wedge (y \notin A)$,	for some element y	7, 6, Conj.
9.	$A \subset C$		4, 8, Def. $\subset$

8. (a) false (b) false (c) false
(d) false (e) false (f) true

9. (a) false (b) false (c) true (d) true (e) false
10. (a) false (b) true (c) false (d) true

11. We shall prove this by mathematical induction on n. Since

$$C(1, r) = \begin{cases} 1 & \text{if } r = 0 \text{ or } 1 \\ 0 & \text{otherwise} \end{cases}$$

the assertion is true for $n = 1$. Suppose that given a set with k elements, then for any integer r, there are exactly $C(k, r)$ subsets with r elements (induction hypothesis). Now consider an arbitrary set $A = \{a_1, a_2, \ldots, a_{k+1}\}$ with $k+1$ elements. The subsets of A with exactly r elements are those subsets of $A - \{a_{k+1}\} = \{a_1, a_2, \ldots, a_k\}$ with r elements, and those obtained by joining the element a_{k+1} to each of the subsets of $A - \{a_{k+1}\}$ with $r-1$ elements. Therefore, by the induction hypothesis and Definition 7, there are exactly $C(k, r) + C(k, r-1) = C(k+1, r)$ subsets of A with r elements. Hence, the proof is complete by the principle of mathematical induction.

EXERCISE 2.2

5. $\varnothing, \{x\}, \{\{y, z\}\}, \{x, \{y, z\}\}$
6. No, $\mathscr{P}(\varnothing) = \{\varnothing\}$ has one element.
7. Two elements.
12. Observe that a set $X \in \mathscr{P}(A : B)$ if and only if there exists a set $Y \in \mathscr{P}(A - B)$ such that $X = B \cup Y$.

(a) Therefore, the number of elements in $\mathscr{P}(A : B)$ is the same as the number of elements of the power set $\mathscr{P}(A - B)$, that is, 2^{n-m}.
(b) If $B = \varnothing$, then $\mathscr{P}(A : B) = \mathscr{P}(A)$, which has $2^{n-0} = 2^n$ elements.

EXERCISE 2.3

1. It is clear the $A \cup B \supseteq B$. To complete the proof, it remains to be shown that $A \cup B \subseteq B$:

$$
\begin{aligned}
x \in A \cup B &\equiv (x \in A) \vee (x \in B) && \text{Def. } \cup \\
&\Rightarrow (x \in B) \vee (x \in B) && A \subseteq B \\
&\equiv x \in B && \text{Idemp.}
\end{aligned}
$$

Hence $A \cup B \subseteq B$ and consequently $A \cup B = B$.

6. (a)

1.	$(x \in A) \to (x \in C)$	(Hyp. $A \subseteq C$)
2.	$(x \in B) \to (x \in C)$ / $\therefore A \cup B \subseteq C$	(Hyp. $B \subseteq C$/Concl.)
3.	$(x \in A) \vee (x \in B) \to (x \in C) \vee (x \in C)$	1, 2, C.D.
4.	$(x \in A) \vee (x \in B) \to (x \in C)$	3, Idemp.
5.	$(x \in A \cup B) \to (x \in C)$	4, Def. $\cup$
6.	$A \cup B \subseteq C$	5, Def. $\subseteq$

7. (i)
$$
\begin{aligned}
C &\subseteq (A \cap B) \cup C \\
&= A \cap (B \cup C) && \text{(Hyp.)} \\
&\subseteq A
\end{aligned}
$$
(ii) Let $C \subseteq A$. Then
$$
\begin{aligned}
(A \cap B) \cup C &= (A \cup C) \cap (B \cup C) && \text{Dist., Com.} \\
&= A \cap (B \cup C) && \text{Prob. 1}
\end{aligned}
$$

11. Since $A \subseteq C$, by Problem 10 we have $A \cup B \subseteq C \cup B$. Since $B \subseteq D$, by Problem 10 again, $C \cup B = B \cup C \subseteq D \cup C = C \cup D$. Thus, $A \cup B \subseteq C \cup D$.

EXERCISE 2.4

1.
$$
\begin{aligned}
A - (B \cap A) &= A \cap (B \cap A)' && \text{Example 5} \\
&= A \cap (B' \cup A') && \text{De Morgan's Th.} \\
&= (A \cap B') \cup (A \cap A') && \text{Dist.} \\
&= (A \cap B') \cup \varnothing && \text{Th. 5(c)} \\
&= A - B && \text{Example 5, Th. 4(a)}
\end{aligned}
$$

3. Let $B \subseteq A'$. For any element x, if $x \in B$ then $x \in A'$. That is $x \notin A$ if $x \in B$. Hence $x \notin A \cap B$ for any x. Therefore $A \cap B = \varnothing$.

Conversily, if $A \cap B = \varnothing$, then $x \in B$ implies $x \notin A$. That is, $(x \in B) \rightarrow (x \in A')$ is true. Thus, $B \subseteq A'$.

6. (i) If $(A - B) \cup B = A$, then $B \subseteq (A - B) \cup B = A$.
   (ii) If $B \subseteq A$, then

$$\begin{aligned}
(A - B) \cup B &= (A \cap B') \cup B && \text{Example 5} \\
&= B \cup (A \cap B') && \text{Com.} \\
&= (B \cup A) \cap (B \cup B') && \text{Dist.} \\
&= A \cap (B \cup B') && \text{(Hyp. } B \subseteq A\text{),} \\
& && \text{Prob. 1 of Ex. 2.3} \\
&= A \cap U && \text{Th. 5(c)} \\
&= A && A \subseteq U
\end{aligned}$$

8. (a)

$$\begin{aligned}
(A - C) \cup (B - C) &= (A \cap C') \cup (B \cap C') && \text{Example 5} \\
&= (C' \cap A) \cup (C' \cap B) && \text{Com.} \\
&= C' \cap (A \cup B) && \text{Dist.} \\
&= (A \cup B) - C && \text{Com., Example 5}
\end{aligned}$$

20.

$$\begin{aligned}
C - B &= (A \cup B) - B && \text{Hyp. } C = A \cup B \\
&= (A \cup B) \cap B' && \text{Example 5} \\
&= (A \cap B') \cup (B \cap B') && \text{Com. Dist.} \\
&= (A - B) \cup \varnothing && \text{Example 5, Th. 5(c)} \\
&= A && \text{Hyp. } A \cap B = \varnothing
\end{aligned}$$

21. $(A \cup B) - (A \cap B)$

$$\begin{aligned}
&= (A \cup B) \cap (A \cap B)' && \text{Example 5} \\
&= (A \cup B) \cap (A' \cup B') && \text{De Morgan's Th.} \\
&= [(A \cup B) \cap A'] \cup [(A \cup B) \cap B'] && \text{Dist.} \\
&= [(A \cap A') \cup (B \cap A')] \cup [(A \cap B') \cup (B \cap B')] && \text{Com., Dist.} \\
&= [\varnothing \cup (B \cap A')] \cup [(A \cap B') \cup \varnothing] && \text{Th. 5(c)} \\
&= (B - A) \cup (A - B) && \text{Example 5} \\
&= (A - B) \cup (B - A) && \text{Com.}
\end{aligned}$$

22. By Example 5, we have

$$B \oplus C = (B - C) \cup (C - B) = (B \cap C') \cup (C \cup B').$$

We shall only prove Part (c) here, the proof of the rest are simple. We have, by Def. of $\oplus$ and De M.,

$$\begin{aligned}
A \oplus (B \oplus C) &= A \oplus [(B \cap C') \cup (C \cap B')] \\
&= \{A \cap [(B \cap C') \cup (C \cap B')]'\} \cup \{A' \cap [(B \cap C') \cup (C \cap B')]\} \\
&= [A \cap (B \cap C')' \cap (C \cap B')'] \cup [A' \cap B \cap C'] \cup [A' \cap B' \cap C] \\
&= [A \cap (B' \cup C) \cap (B \cup C')] \cup [A' \cap B \cap C'] \cup [A' \cap B' \cap C] \\
&= [A \cap B' \cap C'] \cup [A \cap B \cap C] \cup [A' \cap B \cap C'] \cup [A' \cap B' \cap C] \\
&= [A \cap B' \cap C'] \cup [A' \cap B \cap C'] \cup [A \cap B \cap C] \cup [A' \cap B' \cap C]
\end{aligned}$$

$$\begin{aligned}
&= \{[(A \cap B') \cup (A' \cap B)] \cap C'\} \cup \{[(A \cap B) \cup (A' \cap B')] \cap C\} \\
&= [(A \oplus B) \cap C'] \cup [(A' \cup B) \cap (A \cup B') \cap C] \\
&= [(A \oplus B) \cap C'] \cup \{[(A \cap B') \cup (A' \cap B')]' \cap C\} \\
&= [(A \oplus B) \cap C'] \cup [(A \oplus B)' \cap C] \\
&= (A \oplus B) \oplus C
\end{aligned}$$

EXERCISE 2.6

2. (a) $\{0\}$ (b) $[0, 1]$ (c) $[0, 1/99]$ 3. (a) $(-1, 1)$ (b) $\{0\}$

4. (a) $\{x \in \mathbf{R} \mid x > 1\}$ (b) $\mathbf{R}$
(c) $\{x \in \mathbf{R} \mid x \geq 1\}$ (c) $\mathbf{R}$

8.
(a)

$$\begin{aligned}
\left(\bigcup_{i=1}^{m} A_i\right) \cap \left(\bigcup_{j=1}^{n} B_j\right) &= \bigcup_{j=1}^{n}\left(\bigcup_{i=1}^{m} A_i\right) \cap B_j && \text{Th. 9} \\
&= \bigcup_{j=1}^{n}\left[B_j \cap \left(\bigcup_{i=1}^{m} A_i\right)\right] && \text{Com.} \\
&= \bigcup_{j=1}^{n}\left[\bigcup_{i=1}^{m} (B_j \cap A_i)\right] && \text{Th. 9}
\end{aligned}$$

EXERCISE 3.1

2. $A = \varnothing$ or $B = \varnothing$ or $A = B$

9. $(x, y) \in (A \times C) \cap (B \times D)$

$$\begin{aligned}
&\equiv [(x, y) \in A \times C] \wedge [(x, y) \in B \times D] && \text{Def. } \cap \\
&\equiv [(x \in A) \wedge (y \in C)] \wedge [(x \in B) \wedge (y \in D)] && \text{Def. of } \times \\
&\equiv [(x \in A) \wedge (x \in B)] \wedge [(y \in C) \wedge (y \in D)] && \text{Assoc., Com.} \\
&\equiv (x \in A \cap B) \wedge (y \in C \cap D) && \text{Def. } \cap \\
&\equiv (x, y) \in (A \cap B) \times (C \cap D) && \text{Def. of } \times
\end{aligned}$$

Hence, $(A \times C) \cap (B \times D) = (A \cap B) \times (C \cap D)$.

21. (i) If $a = c$ and $b = d$, then $\{a\} = \{c\}$ and $\{a, b\} = \{c, d\}$. Consequently, $\{\{a\}, \{a, b\}\} = \{\{c\}, \{c, d\}\}$, or $(a, b) = (c, d)$.
(ii) Suppose $\{\{a\}, \{a, b\}\} = \{\{c\}, \{c, d\}\}$. If $a = b$, then the ordered pair (a, b) is the same as the singleton set $\{\{a\}\}$. Since $(a, b) = (c, d)$, we have $\{\{c\}, \{c, d\}\} = \{\{a\}\}$. Consequently, $\{c\} = \{c, d\} = \{a\}$ and hence $a = b = c = d$. If $a \neq b$ then both (a, b) and (c, d) contain exactly one singleton set, $\{a\}$ and $\{c\}$ respectively. Hence, $a = c$. The sets (a, b) and (c, d) also contain exactly one "doubleton set," $\{a, b\}$ and $\{c, d\}$ respectively, so that $\{a, b\} = \{c, d\}$. Hence, $b \in \{c, d\}$, so that $b = d$. For if $b = c$ then since $a = c$, we will have $a = b$, a contradiction.

EXERCISE 3.2

3. (a)

$y \in \mathrm{Im}(\mathscr{R}) \equiv (x, y) \in \mathscr{R}$	for some	$x \in A$	Def. of Im
$\equiv (y, x) \in \mathscr{R}^{-1}$	for some	$x \in A$	Def. of $\mathscr{R}^{-1}$
$\equiv y \in \mathrm{Dom}(\mathscr{R}^{-1})$			Def. of Dom

8. 2^{n^2} 9. 2^{mn}

15. Hint: Use Problem 7(b).

EXERCISE 3.3

4.

$(x, y) \in X/\mathfrak{I} \equiv x \in A$ and $y \in A$	for some	$A \in \mathfrak{I}$	Def. $X/\mathfrak{I}$
$\equiv (x, y) \in A \times A$	for some	$A \in \mathfrak{I}$	Def. of $\times$
$\equiv (x, y) \in \bigcup_{A \in \mathfrak{I}} A \times A$			Def. $\cup$

11. $(x, y) \in X/(X/\mathscr{E})$

$\equiv (x, y) \in A \times A$	for some	$A \in X/\mathscr{E}$	Prob. 4
$\equiv (x, y) \in (c/\mathscr{E}) \times (c/\mathscr{E})$	for some	$c \in X$	Def. $X/\mathscr{E}$
$\equiv (x, y) \in \mathscr{E}$			Th. 3(b)

EXERCISE 3.4

6. (a) $[5, +\infty)$ (b) $[-1, 1]$ (c) $\mathbf{R}$ 11. $2^3, n^m$ 12. $2, n$

14. Let $A = \mathrm{Dom}(g)$ and B be any subset of Y containing $\mathrm{Im}(g)$. Show that $g: A \to B$ is a function.

15. Let $(x, y) \in f$. Since f is reflexive, $(x, x) \in f$ so that we must have $y = x$ because f is a function. That is, $f(x) = x$ for all $x \in X$, or $f: X \to X$ is the identity function $I_X : X \to X$.

16. $f(x) = 1 - x$ for all x in $[0, 1]$

17. Let x be any element of X. Since $(x, f(x)) \in f \subseteq g$, we have $(x, f(x)) \in g$. That is, $g(x) = f(x)\ \forall\, x \in X$. By Theorem 7, $f = g$.

EXERCISE 3.5

4. (a)

$x \in A \Rightarrow f(x) \in f(A)$	Def. 9(a)
$\Rightarrow x \in f^{-1}(f(A))$	Def. 9(b)

Therefore, $A \subseteq f^{-1}(f(A))$.

(b) $y \in f(f^{-1}(B)) \Rightarrow \exists\, x \in f^{-1}(B)$ such that $y = f(x)$. Since $x \in f^{-1}(B)$, we have $y = f(x) \in B$. Therefore, $f(f^{-1}(B)) \subseteq \mathbf{B}$.

8. $x \in p_X(\mathscr{R}) \Leftrightarrow x = p_X(x, y)$ for some $(x, y) \in \mathscr{R}$

$\Leftrightarrow x \in \mathrm{Dom}(\mathscr{R})$

Thus, $p_X(\mathscr{R}) = \mathrm{Dom}(\mathscr{R})$. Similarly, $p_Y(\mathscr{R}) = \mathrm{Im}(\mathscr{R})$.

9. (a) $y \in f(A \cap f^{-1}(B))$

$$\begin{aligned} &\equiv [y = f(x) \text{ for some } x \in A \cap f^{-1}(B)] && \text{Def. 9(a)} \\ &\equiv [y = f(x) \text{ for some } x \in A \text{ and } x \in f^{-1}(B)] && \text{Def. } \cap \\ &\equiv y \in f(A) \quad \text{and} \quad y \in B && \text{Def. 9} \\ &\equiv y \in f(A) \cap B && \text{Def. } \cap \end{aligned}$$

Hence, $f(A \cap f^{-1}(B)) = f(A) \cap B$.

(b) Substituting X for A in (a) above, we have $f(X \cap f^{-1}(B)) = f(X) \cap B$. Since $X \cap f^{-1}(B) = f^{-1}(B)$, the last equality may be rewritten as

$$f(f^{-1}(B)) = f(X) \cap B$$

EXERCISE 3.6

7. The X-projection $p_X : X \times Y \to X$ is injective, if Y is a singleton set.

11. $m!$

12. (a) It is always true that $f^{-1}(f(A)) \supseteq A$. [See Problem 4(a), Exercise 3.5.] Therefore, it is sufficient now to show $f^{-1}(f(A)) \subseteq A$: for any $x \in f^{-1}(f(A))$, we have $f(x) \in f(A)$. Consequently, $f(x) = f(x')$ for some $x' \in A$. Since f is injective, we must have $x = x'$; whence, $x \in A$, and thus $f^{-1}(f(A)) \subseteq A$.

16. To show f is symmetric, we show $y = f(x)$ implies $x = f(y)$. Let $y = f(x)$. Then

$$x = f(f(x)) = f(y)$$

18. As a consequence of Theorem 13, for $\Gamma = \{1, 2\}$, $A_1 = A$ and $A_2 = B$, we have $f(A \cap B) = f(A) \cap f(B)$.

To prove the converse, let $f(A \cap B) = f(A) \cap f(B)$ for all subsets A and B of X. For any elements a and b of X, $a \neq b$, we choose $A = \{a\}$ and $B = \{b\}$, then $A \cap B = \varnothing$ and hence $\{f(a)\} \cap \{f(b)\} = f(A) \cap f(B) = f(A \cap B) = f(\varnothing) = \varnothing$. Thus $f(a) \neq f(b)$ and hence f is injective.

19. Let $f(A - B) = f(A) - f(B)$ for all subsets A and B of X. Then for any elements a and b of X, $a \neq b$, we choose $A = \{a\}$ and $B = \{b\}$ so that $A - B \neq \varnothing$, and hence $\{f(a)\} - \{f(b)\} = f(A) - f(B) = f(A - B) \neq \varnothing$. Therefore, $f(a) \neq f(b)$ and hence f is injective.

20. yes.

EXERCISE 3.7

6. To show that f is injective, let $f(x) = f(x')$ for any x and x' in X. Then using $g \circ f = I_X$, we have $x = g \circ f(x) = g \circ f(x') = x'$. Therefore, f is injec-

tive. To show that f is surjective, we use $f \circ h = I_Y$ and the following observation:

$$f(X) \supseteq f(h(Y)) = I_Y(Y) = Y, \qquad \text{because} \qquad h(Y) \subseteq X$$

Therefore, $f(X) = Y$ and f is surjective. Thus, f is a bijection. Next observe that, using the results of Problems 4 and 5, we have

$$g = g \circ I_Y = g \circ (f \circ f^{-1}) = (g \circ f) \circ f^{-1} = I_X \circ f^{-1} = f^{-1}$$

and

$$h = I_X \circ h = (f^{-1} \circ f) \circ h = f^{-1} \circ (f \circ h) = f^{-1} \circ I_Y = f^{-1}$$

12. (i) $g \circ f : X \to Z$ is injective: Let $g \circ f(x) = g \circ f(x')$ for some x and x' in X. Then since g is injective, and $g(f(x)) = g(f(x'))$, we have $f(x) = f(x')$. Then f is injective implies that $x = x'$. This proves that $g \circ f$ is injective.

(ii) To show $g \circ f$ is surjective, observe that

$$\begin{aligned} g \circ f(X) &= g(f(X)) & \\ &= g(Y) & f \text{ is surjective} \\ &= Z & g \text{ is surjective} \end{aligned}$$

Therefore, $g \circ f$ is surjective.

(iii) Now $g \circ f : X \to Z$ is a bijection. Observe that

$$\begin{aligned} (g \circ f) \circ (f^{-1} \circ g^{-1}) &= g \circ (f \circ f^{-1}) \circ g^{-1} & \text{Th. 15} \\ &= (g \circ I_Y) \circ g^{-1} & \text{Prob. 5, Th.15} \\ &= g \circ g^{-1} & \text{Prob. 4} \\ &= I_Z & \text{Prob. 5} \end{aligned}$$

and

$$\begin{aligned} (f^{-1} \circ g^{-1}) \circ (g \circ f) &= f^{-1} \circ (g^{-1} \circ g) \circ f & \text{Th. 15} \\ &= f^{-1} \circ (I_Y \circ f) & \text{Prob. 5, Th. 15} \\ &= f^{-1} \circ f & \text{Prob. 4} \\ &= I_X & \text{Prob. 5} \end{aligned}$$

Therefore, by the result of Problem 6, we have

$$f^{-1} \circ g^{-1} = (g \circ f)^{-1}$$

EXERCISE 4.1

4. If there were a Boolean algebra with exactly three elements, 0, 1 and a, then there would be three possibilities for a':

(i) $a' = 0$ then $a = (a')' = 0' = 1$

(ii) $a' = 1$ then $a = (a')' = 1' = 0$, by Th. 4.

(iii) $a' = a$ then $a = a*a = a*a' = 0$.

All the three possibilities lead to contradictions.

5. If $a' = a$, then
$1 = a + a' = a + a = a = a*a = a*a' = 0$, which contradicts Def. 1(d).

9. (a) $a + (a*b) = (a*1) + (a*b)$
$= a*(1 + b)$
$= a*1$ by Th. 5.
$= a$

12. If $a + b = 0$, then by Prob. 9(b),
$$a = a*(a + b) = a*0 = 0 \quad \text{by Th. 5}$$
Similarly, $b = 0$.

15. (a) ⇒ (b): $a' + b = a' + (a + b) = (a' + a) + b = 1 + b = 1$, by Th. 5
(b) ⇒ (c): $a = a*1 = a*(a' + b) = (a*a') + (a*b)$
$= 0 + (a*b) = a*b$
(c) ⇒ (d): $a*b' = (a*b)*b' = a*(b*b') = a*0 = 0$
(d) ⇒ (a): $b = b + 0 = b + (a*b') = (b + a)*(b + b')$
$= (b + a)*1 = b + a = a + a$

EXERCISE 4.2

7. $x'yz + x'yz' + xy'z + xy'z' = x'y(z + z') + xy'(z + z')$
$= x'y \cdot 1 + xy' \cdot 1$
$= x'y + xy'$

11. $xy + zw + x + (x + z' + w')' = (x + xy) + zw + x'zw$ by De M. Laws
$= x + zw(1 + x')$ by Absorption Laws
$= x + zw$ by Theorem 5

16. $f(x, y, z) = xx' + yy' + zz'$

EXERCISE 4.3

3. $(x \leftrightarrow y) \leftrightarrow z = (x \leftrightarrow y)z + (x \leftrightarrow y)'z'$
$= (xy + x'y')z + (xy + x'y')'z'$
$= (xy + x'y')z + (x' + y')(x + y)z'$
$= xyz + x'y'z + x'yz' + xy'z'$

$x \leftrightarrow (y \leftrightarrow z) = x(y \leftrightarrow z) + x'(y \leftrightarrow z)'$
$= x(yz + y'z') + x'(yz + y'z')$
$= xyz + xy'z' + x'y'z'$
Hence, $(x \leftrightarrow y) \leftrightarrow z = x \leftrightarrow (y \leftrightarrow z)$.

4. $x \oplus (y \oplus z) = x \oplus (yz' + y'z)$
$= x(yz' + y'z)' + x'(yz' + y'z)$
$= x(yz')'(y'z)' + x'yz' + x'y'z$
$= x(y' + z)(y + z') + x'yz' + x'y'z$
$= xy'z' + xyz + x'yz' + x'y'z$

$$= (xy' + x'y)z' + (xy + x'y')z$$
$$= (x \oplus y)z' + (x' + y)(x + y')z$$
$$= (x \oplus y)z' + (xy' + x'y)'z$$
$$= (x \oplus y)z' + (x \oplus y)'z = (x \oplus y) \oplus z$$

EXERCISE 4.4

3. This was proved in the passage of the proof of Problem 4, Exercise 4.3

4. $C = xyz + xyz' + xy'z + x'yz$
$= xyz + xyz + xyz + xyz' + xy'z + x'yz$, by Th. 3
$= (xyz + xyz') + (xyz + xy'z) + (xyz + x'yz)$
$= xy(z + z') + xz(y + y') + (x + x')yz$
$= xy \cdot 1 + xz \cdot 1 + 1 \cdot yz$
$= xy + xz + yz;$

$C = xyz + xyz' + xy'z + x'yz$
$= xy(z + z') + (xy' + x'y)z$
$= xy + (xy' + x'y)z$

5. (a) $B = x'y$ and $D = (x' + y')(x + y)$
(b) $B = x'y$ and $D = x'y + xy' = x \oplus y$

6. (b)

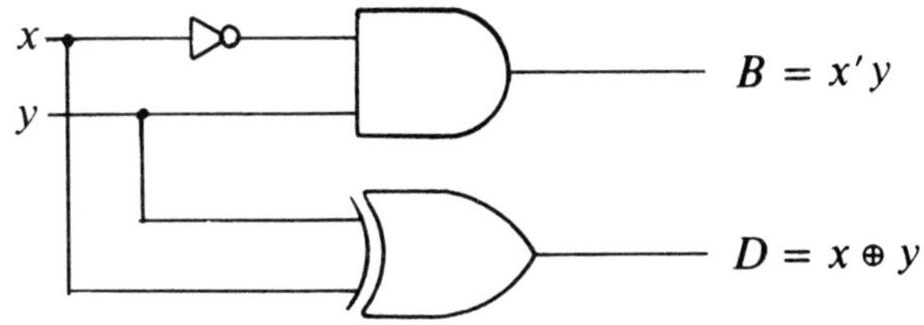

7. (a) $B = x'y'z + x'yz' + x'yz + xyz$, and
$D = x'y'z + x'yz' + xy'z' + xyz$

Notice that the function D happens to be identical to the function S of Problem 3.

(b) $B = x'(y'z + yz') + (x' + x)yz$
$= x'(y \oplus z) + yz$
$D = S$ of Prob. 3
$= (x \oplus y) \oplus z = x \oplus (y \oplus z)$, by Prob. 4, Ex. 4.3.

9.

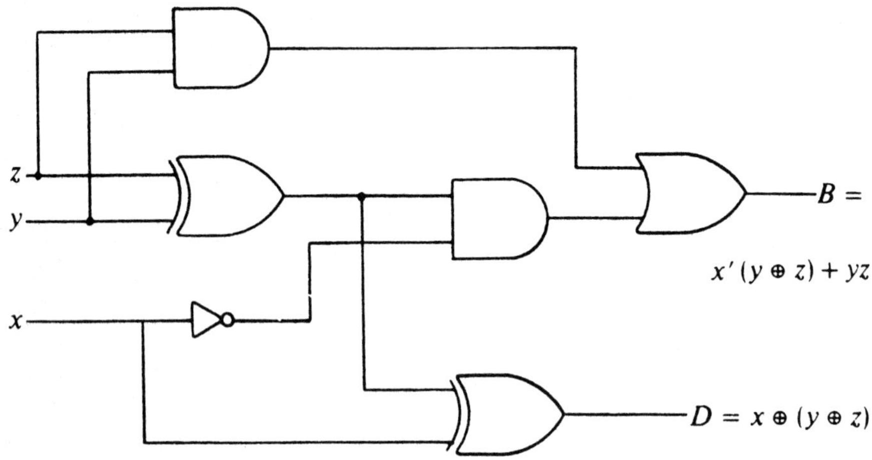

EXERCISE 5.1

2. Suppose on the contrary that Y is infinite. Then since $g^{-1}: Y \to X$ is a one-to-one correspondence, by Theorem 2, the set X must be infinite, a contradiction. Therefore, Y is finite.

4. Let $f: A \to A$ be an injection such that $f(A) \neq A$. Then the function $g: A \times A \to A \times A$ defined by $g(x, y) = (f(x), f(y))$ is injective, and $g(A \times A) \neq A \times A$. Hence, $A \times A$ is infinite.

15. If $A \oplus B = (A - B) \cup (B - A)$ is infinite, then by Problem 11 either $A - B$ or $B - A$ is infinite. Assuming $A - B$ is infinite then the set A, as a superset of the infinite set $A - B$, is infinite. The converse is not true, for if $A = B$ is an infinite set then $A \oplus B = A \oplus A = \varnothing$ is finite.

EXERCISE 5.2

4. Let $f: X - Y \sim Y - X$. Then consider the function $g: X \to Y$ defined by

$$g(x) = \begin{cases} f(x) & \text{if } x \in X - Y \\ x & \text{if } x \in X \cap Y \end{cases}$$

Since $X - Y = X - (X \cap Y)$ and $Y - X = Y - (X \cap Y)$. the function $g: X \to Y$ is bijective. Hence, $X \sim Y$.

5. Let $f_\gamma : X_\gamma \sim Y_\gamma$ for each $\gamma \in \Gamma$. Define the function $f: \bigcup_{\gamma\in\Gamma} X_\gamma \rightarrow \bigcup_{\gamma\in\Gamma} Y_\gamma$ by $f(x) = f_\gamma(x)$ if $x \in X_\gamma$. Then since $\{X_\gamma \mid \gamma \in \Gamma\}$ and $\{Y_\gamma \mid \gamma \in \Gamma\}$ are families of disjoint sets, f is a well-defined bijection.
9. The function $f: \mathscr{P}(A) \rightarrow 2^A$ which takes each subset B of A to the characteristic function $\chi_B : A \rightarrow \{0, 1\}$ of B is clearly a bijection.
10. (i) We first prove that h is injective: Let $x, y \in Y$ such that $h(x) = h(y)$, that is

$$\{1, 2, \ldots, g(x)\} \cap g(Y) = \{1, 2, \ldots, g(y)\} \cap g(Y)\} \qquad (*)$$

We shall show $g(x) = g(y)$ and hence $x = y$. For, if otherwise, we may assume without loss of generality that $g(x) < g(y)$, then the number of elements in the left hand side set of (*), call it LHS, is at least one less than the number of elements of the right hand side set of (*), call it RHS; because under $g(x) < g(y)$ the LHS is a subset of RHS, and the element $g(y)$ in RHS is not in LHS. This contradiction shows $g(x) = g(y)$ and hence $x = y$.
(ii) We next show that h is surjective: Suppose otherwise, then there exists $n_0 \in \mathbf{N}$ such that $n_0 \notin h(Y)$. Without loss of generality, let n_0 be the smallest such natural number. Then there exists $y_0 \in Y$ such that $h(y_0) = n_0 - 1$ (it is easy to argue that $n_0 \neq 1$.) Let $y_1 =$ the smallest of $\{g(z) \mid g(z) > g(y_0), z \in Y\}$. It follows that

$$\begin{aligned} h(y_1) &= \text{the number of elements in } \{1, 2, \ldots, g(y_1)\} \cap g(Y) \qquad (**) \\ &= (n_0 - 1) + 1 = n_0 \end{aligned}$$

because,

$$\{1, 2, \ldots, g(y_1)\} \cap g(Y) = [\{1, 2, \ldots, g(y_0)\} \cap g(Y)] \cup \{g(y_1)\}$$

The result of (**) contradicts the earlier supposition that $n_0 \notin h(Y)$. Therefore, h is surjective.

EXERCISE 5.3

4. Let $f: A \sim \mathbf{N}$ and $g: B \sim \mathbf{N}$. Then the function $h: A \times B \rightarrow \mathbf{N} \times \mathbf{N}$ defined by $h(x, y) = (f(x), g(y))$ is a bijection. Hence, $A \times B \sim \mathbf{N} \times \mathbf{N}$. Consequently, by Theorem 10, $A \times B$ is denumerable.

5. Let us express each rational number uniquely as p/q, where $p \in \mathbf{Z}$, $q \in \mathbf{N}$, and the greatest common divisor of p and q is 1. Then the function $f: \mathbf{Q} \rightarrow \mathbf{Z} \times \mathbf{N}$ defined by $f(p/q) = (p, q)$ is an injection. Clearly, $\mathbf{Z} \times \{1\} \subseteq f(\mathbf{Q}) \subseteq \mathbf{N} \times \mathbf{N}$. Therefore, by Theorem 8, $f(\mathbf{Q})$ is denumerable and hence so is $\mathbf{Q}$.

6. Let $\mathscr{C}$ be the set of all circles in the Cartesian plane having rational radii and centers at points having both coordinates rational. Consider the function $f: \mathscr{C} \rightarrow \mathbf{Q} \times \mathbf{Q} \times \mathbf{Q}$ defined by $f(c) = (x, y, z)$ where (x, y) is the center and z

the radius of the circle $c \in \mathscr{C}$. Then clearly f is an injection. Observe that, by Example 5, $\mathbf{Q} \sim \mathbf{N}$ and hence $\mathbf{Q} \times \mathbf{Q} \times \mathbf{Q} \sim \mathbf{N} \times \mathbf{N} \times \mathbf{N} \sim \mathbf{N} \times \mathbf{N} \sim \mathbf{N}$, by Theorem 10. Now $f(\mathscr{C})$ is an infinite subset of the denumerable set $\mathbf{Q} \times \mathbf{Q} \times \mathbf{Q}$, so by Theorem 8, $f(\mathscr{C})$ is denumerable. Hence, $\mathscr{C}$ is denumerable.

7. Let $A_1 = B_1$ and $A_{k+1} = B_{k+1} - \bigcup_{j=1}^{k} A_j$ for each $k \in \mathbf{N}$. Then $\{A_k \mid k \in \mathbf{N}\}$ is a denumerable family of disjoint countable sets. Furthermore, $\bigcup_{k \in \mathbf{N}} A_k = \bigcup_{k \in \mathbf{N}} B_k$ and $A_1 = B_1$ is denumerable. By using the corollary to Theorem 10, $\bigcup_{k \in \mathbf{N}} A_k$ may be shown to be denumerable, and hence so is $\bigcup_{k \in \mathbf{N}} B_k$.

10. For each $k \in \mathbf{N}$, let A_k be the set of all polynomials $a_0 x^k + a_1 x^{k-1} + \cdots + a_k$ with integer coefficients and $a_0 \neq 0$. Then each A_k is denumerable, and hence by the Corollary to Theorem 10, $\bigcup_{k \in \mathbf{N}} A_k$ is denumerable.

12. Let A be a denumerable set and let, for each $k \in \mathbf{N}$, A_k be the set of all finite subsets of A having precisely k elements. Then for each $k \in \mathbf{N}$
$$\begin{aligned} A_k &\sim \mathbf{N} \times \mathbf{N} \times \cdots \times \mathbf{N} \ (k \text{ factors}) \\ &\sim \mathbf{N} \times \mathbf{N} \times \cdots \times \mathbf{N} \ (k-1 \text{ factors}) \\ &\sim \mathbf{N} \end{aligned}$$
By the Corollary to Theorem 10, the set $\bigcup_{k \in \mathbf{N}} A_k$ of all nonempty finite subsets of A is denumerable.

EXERCISE 5.4

4. Suppose on the contrary that the set of all irrational numbers between 0 and 1 is denumerable. since the set of all rational numbers between 0 and 1 is denumerable, the union of these two sets, which constitutes the set of real numbers between 0 and 1, must be denumerable. This contradicts Theorem 12. Therefore, the set of all irrational numbers between 0 and 1 is nondenumerable.

6. $S^1 \sim [0, 2\pi) \sim \mathbf{R}$.

11. If $A \oplus B = (A - B) \cup (B - A)$ is nondenumerable, then by Problem 10 either $A - B$ or $B - A$ is nondenumerable. Assume $A - B$ is nondenumerable, the its superset A is nondenumerable. The converse is not true.

EXERCISE 6.1

2. $0 = \operatorname{card} \varnothing$, $\operatorname{card} \mathbf{N}$, and $\operatorname{card} \mathbf{R}$

5. yes

EXERCISE 6.2

2. Let a be a transfinite cardinal number and let A be a set such that $\text{card}A = a$. Then the set A is an infinite set which, by Theorem 11 of Chapter 5, contains a denumerable subset A. That is, $\mathbf{N} \sim B \subseteq A$, which shows that $\text{card}\mathbf{N} \leq a$.

7. Consider the sets B and C. Since $B \sim B \subseteq C$ and $C \sim A \subseteq B$, by the Schröder-Bernstein Theorem $C \sim B$. It follows from $A \sim C$ and $C \sim B$ that $A \sim B$.

EXERCISE 6.3

2. Let $f: A \sim B$. The function $f: A \to B$ induces the function $f^*: \mathscr{P}(A) \to \mathscr{P}(B)$ defined by $f^*(X) = f(X)$ for all $X \in \mathscr{P}(A)$. Since f is bijective, so is f^*. Therefore, $\mathscr{P}(A) \sim \mathscr{P}(B)$.

3. Suppose on the contrary that there is a denumerable set A whose power set $\mathscr{P}(A)$ is denumerable. Then $\text{card}A = \text{card}\,\mathscr{P}(A)$, which is a contradiction to Cantor's Theorem (Theorem 2).

5. If there were a set $\mathscr{U}$ of all sets, then it would be true that card $\mathscr{U} > A$ for any set A. But according to Cantor's Theorem card $\mathscr{U} <$ card $\mathscr{P}(\mathscr{U})$ which would be a contradiction. Therefore, the set of all sets does not exist.

EXERCISE 6.4

5. Since $\mathbf{R} \sim (0, 1) \sim (1, 2)$ and $(0, 1) \subseteq (0, 1) \cup (1, 2) \subseteq \mathbf{R}$, by the Schröder-Bernstein Theorem we have $(0, 1) \cup (1,2) \sim \mathbf{R}$ which shows that $c + c = c$.

8. Counterexample: $2 + \text{card}\mathbf{N} = 3 + \text{card}\mathbf{N}$, but $2 \neq 3$.

EXERCISE 6.5

3. Counterexample: Let $x = 1, y = z = \aleph_0$. Then $x < y$, but $xz = yz = \aleph_0$ (Example 5(c)).

9. Counterexample: $2 \cdot \text{card}\mathbf{N} = 1 \cdot \text{card}\mathbf{N}$, but $2 \neq 1$.

EXERCISE 6.6

5. For any $n \geq 2$, by Theorems 8 and 10 and the fact that $\aleph_0 \aleph_0 = \aleph_0$, we have

$$c = 2^{\aleph_0} \leq n^{\aleph_0} \leq \aleph_0^{\aleph_0} \leq (2^{\aleph_0})^{\aleph_0} = 2^{\aleph_0 \aleph_0} = 2^{\aleph_0} = c$$

Thus, $n^{\aleph_0} = c = \aleph_0^{\aleph_0}$.

8. We have $c \leq \aleph_0 c \leq cc = c$ by Problem 6. Hence, $\aleph_0 c = c$.

EXERCISE 6.7

2. Let $\mathbf{H}$ denote the classic Hilbert space. Then we have

$$c \leq \operatorname{card}\mathbf{H} \leq c^{\aleph_0} = (2^{\aleph_0})^{\aleph_0} = 2^{\aleph_0 \aleph_0} = 2^{\aleph_0} = c$$

by Theorems 8 and 10. Hence, $\operatorname{card}\mathbf{H} = c = \operatorname{card}\mathbf{R}$.

3. The set of lattice points in $\mathbf{R}^{\aleph_0}$ has the cardinal number $\aleph_0^{\aleph_0} = c$, by Problem 5 of Exercise 6.6, and $\operatorname{card}\mathbf{R}^{\aleph_0} = c^{\aleph_0} = c$ as shown in Problem 2 above.

4. The cardinal number of the set of all functions of one real variable which assume only the values 0 and 1 is 2^c, and the cardinal number of the set of all real-valued functions of n variables is c^{c^n}. Observe that, since $c^n = c$ and $\aleph_0 c = c$, we have

$$c^{c^n} = c^c = (2^{\aleph_0})^c = 2^{\aleph_0 c} = 2^c$$

5. From the result of Problem 4 above, we have $\mathfrak{f} = 2^c$. Observe now that

$$\mathfrak{f} \leq \mathfrak{f}^n \leq \mathfrak{f}^{\aleph_0} \leq \mathfrak{f}^c = (2^c)^c = 2^{cc} = 2^c = \mathfrak{f}$$

EXERCISE 7.1

2. We may assume that $B \neq \varnothing$ (for if $B = \varnothing$ then $C = \varnothing$). The set

$$\{f^{-1}(y) \mid y \in B\}$$

forms a partition of A. By the axiom of choice, the set $\{f^{-1}(y) \mid y \in B\}$ has a set C of representatives such that $C \cap f^{-1}(y)$ is a singleton set for each $y \in B$, so that the restriction of f to C, $f \mid C : C \to B$, is bijective. Hence, $\operatorname{card}A \geq \operatorname{card}C = \operatorname{card}B$.

8. Let $g : \{A_\gamma \mid \gamma \in \Gamma\} \to \bigcup_{\gamma \in \Gamma} A_\gamma$ be a choice function, by the axiom of choice. Then there exists a function $f : \Gamma \to \bigcup_{\gamma \in \Gamma} A_\gamma$ defined by $f(\gamma) = g(A_\gamma) \in A_\gamma$ for all $\gamma \in \Gamma$. Hence $\mathbf{P}_{\gamma \in \Gamma} A_\gamma \neq \varnothing$ if $\Gamma \neq \varnothing$.

EXERCISE 7.2

12. Let $\mathscr{T}$ be the set of all totally ordered subsets of A that contain B. $\mathscr{T}$ can be partially ordered by the inclusion, $\subseteq$. Then the same proof for Theorem 2 applies here to conclude that $\mathscr{T}$ has a maximal member C, $B \subseteq C$.

EXERCISE 7.3

2. Let $(A, \leq)$ be a partially ordered set, and let the set $\mathcal{T}$ of all totally ordered subsets of $(A, \leq)$ be partially ordered by $\subseteq$. In order to apply Zorn's lemma to $(\mathcal{T}, \subseteq)$, let $\mathscr{C}$ be a chain in $(\mathcal{T}, \subseteq)$, and let $K = \bigcup_{C \in \mathscr{C}} C$. We will show that $K \in \mathcal{T}$ and hence it is an upper bound for $\mathscr{C}$. Indeed, if $x, y \in K$, then $x \in D$ and $y \in E$ for some $D \in \mathscr{C}$ and $E \in \mathscr{C}$. But $\mathscr{C}$ is a chain in $(\mathcal{T}, \subseteq)$; hence either $D \subseteq E$ or $E \subseteq D$. Assume that $E \subseteq D$; then $x, y \in D$. But D is totally ordered, so that either $x \leq y$ or $y \leq x$. Hence K is a totally ordered subset of $(A, \leq)$, so that $K \in \mathcal{T}$. Now by Zorn's lemma $(\mathcal{T}, \subseteq)$ has a maximal element.

3. Let R be a ring with identity 1, and let the set $\mathscr{A}$ of all proper ideals of R be partially ordered by $\subseteq$. Let $\mathscr{C}$ be a chain in $(\mathscr{A}, \subseteq)$. Then $\bigcup_{I \in \mathscr{C}} I$ is a proper ideal of R because $1 \notin \bigcup_{I \in \mathscr{C}} I$, so that $\bigcup_{I \in \mathscr{C}} I \in \mathscr{A}$ is an upper bound for $\mathscr{C}$. By Zorn's lemma, $(\mathscr{A}, \subseteq)$ has a maximal member.

4. Let V be a vector space, and let the set $\mathscr{A}$ of all linearly independent subsets of vectors in V be partially ordered by $\subseteq$. Let $\mathscr{C}$ be a chain in $(\mathscr{A}, \subseteq)$; then $K = \bigcup_{C \in \mathscr{C}} C$ is clearly linearly independent, and hence $K \in \mathscr{A}$ is an upper bound for $\mathscr{C}$. By Zorn's lemma, $(\mathscr{A}, \subseteq)$ has a maximal member (which forms a basis for V).

8. Let $\mathfrak{F}$ be partially ordered by $\subseteq$, and let $\mathscr{C}$ be a chain in $(\mathfrak{F}, \subseteq)$. We shall show that $\mathscr{C}$ has an upper bound. The natural candidate for this upper bound is $K = \bigcup_{C \in \mathscr{C}} C$. Let $\{x_1, x_2, \ldots, x_n\}$ be any finite subset of K. Then $x_i \in C_i$ for some $C_i \in \mathscr{C}$, $i = 1, 2, \ldots, n$. But, since $\mathscr{C}$ is a chain, there exists $C \in \mathscr{C}$ such that $C_i \subseteq C$ for all $i = 1, 2, \ldots, n$. Whence, $\{x_1, x_2, \ldots, x_n\} \subseteq C$ and hence $\{x_1, x_2, \ldots, x_n\} \in \mathfrak{F}$. Therefore $K \in \mathfrak{F}$ and hence it is an upper bound for $\mathscr{C}$. By Zorn's lemma, $(\mathfrak{F}, \subseteq)$ has a maximal element.

EXERCISE 7.4

2. Since the set $\mathbf{Q}$ is denumerable, there exists a bijection $f: \mathbf{Q} \sim \mathbf{N}$. Define a relation $\leq$ on $\mathbf{Q}$ by declaring that for any p and q in $\mathbf{Q}$, $p \leq q$ if and only if $f(p) \leq f(q)$ in $\mathbf{N}$ under the natural ordering of the natural numbers. Since $(\mathbf{N}, \leq)$ is well ordered, so is $(\mathbf{Q}, \leq)$.

7. If $(A, \leq)$ is well ordered, it cannot contain an infinite strictly decreasing sequence, for this would be a subset without a least element. Conversely, suppose that the totally ordered set $(A, \leq)$ is not well ordered. Then there exists a subset B of A with no least element. Choose any $a_1 \in B$. Since a_1 is not the least element of B, we can choose $a_2 \in B$ such that $a_1 > a_2$. Similarly we can choose $a_3 \in B$ such that $a_2 > a_3$. Continuing, we have an infinite strictly decreasing sequence $a_1 > a_2 > a_3 > \cdots$.

EXERCISE 8.2

7. ω

EXERCISE 8.3

4. Let $\alpha = \omega$, $\beta = 0$, and $\gamma = 1$. Then $\beta + \alpha = \omega = \gamma + \alpha$, but $\beta \neq \gamma$.

6. Use the same example as Problem 4 above: $\beta = 0 \prec 1 = \gamma$, but $\beta+\alpha = \gamma+\alpha$.

EXERCISE 8.4

4. Let $\alpha = \omega$, $\beta = \gamma = 1$. Then $(\beta+\gamma)\alpha = 2\omega = \omega$, but $\beta\alpha + \gamma\alpha = \omega + \omega = \omega 2$. Thus, $(\beta+\gamma)\alpha \neq \beta\alpha+\gamma\alpha$.

8. Let $\alpha = 1$, $\beta = 2$, $\gamma = \omega$. Then $\alpha \prec \beta$ and $\gamma \succ 0$, but $\alpha\gamma = 1\omega = \omega = 2\omega = \beta\gamma$.

EXERCISE 8.5

3. Let X be any set. By the well-ordering principle, the set X can be well ordered. Let $\mathcal{S} = \{\text{ord}(X, \leq) \mid \leq \text{ is a well-order relation of } X\}$. By Theorem 12, the set $\mathcal{S}$ is well ordered, and hence has a unique least element, the initial ordinal for X.

Glossary of Symbols and Abbreviations

$\sim$	not . . .
$\wedge$	and
$\vee$	or
$\rightarrow$	if . . . then
$\leftrightarrow$	if and only if
$\equiv$ ($\Leftrightarrow$)	is equivalent to
$\Rightarrow$	implies
$\forall$	for all (for every)
$\exists$	there exist(s) (for some)
$\therefore$	therefore
$a \in A$	a is an element of A
$a \notin A$	a is not an element of A
$\{x, y, \ldots, z\}$	the set consisting of the elements $x, y, \ldots,$ and z
$\{x \mid p(x)\}$	set of all x such that $p(x)$ is true
$\mathbf{N}$	set of all natural numbers
$\mathbf{Z}$	set of all integers
$\mathbf{Q}$	set of all rational numbers
$\mathbf{R}$	set of all real numbers
$\mathbf{R}_+$	set of all positive real numbers
$A \subseteq B$	A is a subset of B
$A \nsubseteq B$	A is not a subset of B
$A \subset B$	A is a proper subset of B
$A \supseteq B$	A is a superset of B
$\varnothing$	empty set

$A \cap B$	intersection of sets A and B
$A \cup B$	union of sets A and B
$\mathcal{P}(A)$	power set of A (the set of all subsets of A)
$\bigcap_{\gamma \in \Gamma} C_\gamma$	intersection of the sets C_γ where $\gamma \in \Gamma$
$\bigcup_{\gamma \in \Gamma} C_\gamma$	union of the sets C_γ where $\gamma \in \Gamma$
$\bigcap_{k=1}^{n} C_k$	intersection of the sets $C_1, C_2, \ldots, C_n$
$\bigcup_{k=1}^{n} C_k$	union of the sets $C_1, C_2, \ldots, C_n$
$\bigcap_{A \in \mathcal{F}} A$	intersection of the family $\mathcal{F}$ of sets A
$\bigcup_{A \in \mathcal{F}} A$	union of the family $\mathcal{F}$ of sets A
(a, b)	ordered pair of elements a, b
$A \times B$	Cartesian product of sets A and B
$\mathrm{Dom}(\mathcal{R})$	domain of the relation $\mathcal{R}$
$\mathrm{Im}(\mathcal{R})$	image of the relation $\mathcal{R}$
$\mathcal{R}^{-1}$	inverse of the relation $\mathcal{R}$
Δ_X	identity relation on X
$\mathcal{E}$	equivalence relation
$x/\mathcal{E}$	equivalent class determined by x and $\mathcal{E}$
$X/\mathcal{E}$	set of all equivalent classes $x/\mathcal{E}$ where $x \in X$
$\mathcal{T}$	partition of a set
$X/\mathcal{T}$	equivalence relation induced by the partition $\mathcal{T}$ of X
$f: X \to Y$	f is a function from X to Y
$I_X: X \to X$	identity function on X
$\chi_A: X \to \{1, 2\}$	characteristic function where $A \subseteq X$
$f: X \sim Y$	function $f: X \to Y$ is a one-to-one correspondence
2^A	set of all functions from A to $\{1, 2\}$
B^A	set of all functions from A to B
$A \sim B$	A is equipotent to B
$\mathrm{card}\, A$	cardinal number of the set A
$\aleph_0$	cardinal number of the set of natural numbers
c	cardinal number of the set of real numbers
$\mathfrak{f}$	cardinal number of the set of all functions from $\mathbf{R}$ to $\mathbf{R}$
A_x	segment $\{a \in A \mid a < x\}$ of A, where $(A, \leq)$ is a well-ordered set
$(A, \leq) \approx (B, \leq')$	the well-ordered sets $(A, \leq)$ and $(B, \leq')$ are order isomorphic
$\mathrm{ord}(A, \leq)$	ordinal number of $(A, \leq)$
ω	ordinal number of $(\mathbf{N}, \leq)$

Add.	Law of Addition: $p \Rightarrow p \vee q$
Simp.	Laws of Simplification: $p \wedge q \Rightarrow p$, $p \wedge q \Rightarrow q$
D.S.	Law of Double Syllogism: $(p \vee q) \wedge \sim p \Rightarrow q$
D.N.	Law of Double Negations: $\sim(\sim p) \equiv p$

Com.	Commutative Laws: $p \wedge q \equiv q \wedge p$, $p \vee q \equiv q \vee p$
Idemp.	Laws of Idempotencey: $p \wedge p \equiv p$, $p \vee p \equiv p$
Contrap.	Contrapositive Law: $(p \rightarrow q) \equiv (\sim q \rightarrow \sim p)$
De M.	De Morgan's Laws: $\sim(p \wedge q) \equiv \sim p \vee \sim q$, $\sim(p \vee q) \equiv \sim p \wedge \sim q$
Assoc.	Associative Laws: $(p \wedge q) \wedge r \equiv p \wedge (q \wedge r)$ $(p \vee q) \vee r \equiv p \vee (q \vee r)$
Dist.	Distributive Laws: $p \wedge (q \vee r) \equiv (p \wedge q) \vee (p \wedge r)$ $p \vee (q \wedge r) \equiv (p \vee q) \wedge (p \vee r)$
Trans.	Transitive Law: $(p \rightarrow q) \wedge (q \rightarrow r) \Rightarrow (p \rightarrow r)$
C.D.	Constructive Dilemmas: $(p \rightarrow q) \wedge (r \rightarrow s) \Rightarrow (p \vee r \rightarrow q \vee s)$ $(p \rightarrow q) \wedge (r \rightarrow s) \Rightarrow (p \wedge r \rightarrow q \wedge s)$
D.D.	Destructive Dilemmas: $(p \rightarrow q) \wedge (r \rightarrow s) \Rightarrow (\sim q \vee \sim s \rightarrow \sim p \vee \sim r)$ $(p \rightarrow q) \wedge (r \rightarrow s) \Rightarrow (\sim q \wedge s \rightarrow \sim p \wedge \sim r)$
M.P.	Modus Ponens: $(p \rightarrow q) \wedge p \Rightarrow q$
M.T.	Modus Tollens: $(p \rightarrow q) \wedge \sim q \Rightarrow \sim p$
R.A.	Reductio ad Absurdum: $(p \rightarrow q) \Leftrightarrow (p \wedge \sim q \rightarrow q \wedge \sim q)$
Q.N.	Rule of Quantifier Negation: $\sim[(\forall x)(p(x))] \equiv (\exists x)(\sim p(x))$ $\sim[(\exists x)(p(x))] \equiv (\forall x)(\sim p(x))$
Def.	Definition
I.P.	Method of Indirect Proof
Th.	Theorem
Hyp.	Hypothesis
Concl.	Conclusion
t	tautology
c	contradiction

Selected Bibliography

For those who wish to study further in the theory of sets, we recommend the following books:

I. Books of Historical and Philosophical Interest

Benaceraf, Paul, and Hilary Putnam, eds. *Philosophy of Mathematics.* Prentice-Hall, Inc., Englewood Cliffs, N.J., 1964.

Van Heijenoort, Jean, ed. *From Frege to Gödel.* Harvard University Press, Cambridge, Mass., 1967.

Wilder, Raymond L. *Introduction to the Foundations of Mathematics.* 2nd ed. John Wiley & Sons,Inc., New York, 1965.

II. Books Related to Logic

Copi, Irving M. *Symbolic Logic.* 3rd ed. The Macmillan Company, 1976.

Kleene, Stephen C. *Mathematical Logic.* John Wiley & Sons, Inc., New York, 1967.

III. Books Related to Set Theory

Fraenkel, Abraham A. *Set Theory and Logic.* Addison-Wesley Publishing Company, Inc., Reading, Mass., 1966.

Halmos, Parl R. *Naive Set Theorey.* D. Van Nostrand Company, Inc., Princeton, N.J., 1960.

Hayden, Seymour, and John F. Kennison. *Zermelo-Fraenkel Set Theory.* Charles E. Merrill Publishing Company, Columbus, Ohio, 1968.

Monk, J. Donald. *Introduction to Set Theory.* McGraw-Hill, Inc., New York, 1969.

Pinter, Charles C. *Set Theory.* Addison-Wesley Publishing Company, Inc., Reading, Mass., 1971.

Suppes, Patrick. *Axiomatic Set Theory.* D. Van Nostrand Company, Inc., Princeton, N.J., 1960.

IV. Books Related to Boolean Algebra with Applications

Boole, G. *An Introduction of the Law of Thought.* Dover Publications, Inc., New York, 1954.

Mano, M. Morris. *Digital Logic and Computer Design.* Prentice-Hall, Inc., Englewood Cliffs, N.J., 1979.

Index